Information Organization of The Universe and Living Things

Digital Sciences Set

coordinated by
Abdelkhalak El Hami

Volume 3

Information Organization of The Universe and Living Things

Generation of Space, Quantum and Molecular Elements, Coactive Generation of Living Organisms and Multiagent Model

Alain Cardon

ISTE

WILEY

First published 2021 in Great Britain and the United States by ISTE Ltd and John Wiley & Sons, Inc.

ISTE Ltd
27-37 St George's Road
London SW19 4EU
UK

www.iste.co.uk

John Wiley & Sons, Inc.
111 River Street
Hoboken, NJ 07030
USA

www.wiley.com

Library of Congress Control Number: 2021941901

British Library Cataloguing-in-Publication Data
A CIP record for this book is available from the British Library
ISBN 978-1-78630-746-0

MIX
Paper from responsible sources
FSC® C013604

Contents

Introduction

The Universe is a huge expanse of space containing innumerable quantum, molecular and material elements. The physical sciences have discovered space-time and quantum particles, and there exists the possibility of producing scientific theories about the creation and the reasons for the state of the Universe, specifying the generation, the role and the action of all the elements that it contains. We will propose a universe generation model by considering that it is formed in regard to the expansion of an informational substratum possessing great informational energy and applying everywhere a self-organizing law to produce, by emergence, the space and all the material elements in continuous organization.

To understand what the Universe is, we scientifically observe very many elements at multiple scales, compelling us to ask three major fundamental questions and provide precise answers:

– What was the initial state of generation of this universe, a state where there was neither space nor time and where nothing physical existed?

– What is the space of the Universe, how did the space that allows the generation and deployment of all physical elements come to be created?

– What is this organization of the Universe that allowed the creation of atoms, molecules, clouds of physical components, then stars, galaxies and planets with physical and temporal stability, and what is it that allowed the creation of the extraordinary development of life on Earth, up to humans who ask existential questions?

To answer these questions, we can define a truly unifying model for the notions of generation of space, of quantum elements and of all material elements and also of all living organisms. We must posit that there was an empty space available upon the creation of the Universe and then particles in place there, developing and unfolding at random using energy. We need an answer with a conceptual unification of the creation of the space of the Universe, of the quantum particles, of the molecules and of the material aggregates, and we need to define what this energy was which enabled the creation of all that and especially to specify why the physical elements aggregated to constitute such aggregates leading to the suns, to the planets and to life. In fact, it is necessary to make a unifying analysis of the generation of the Universe and not only multiple local analyses, in bottom-up and top-down approaches. It is necessary to define an organizational law that engages in a continuous way the organization of the Universe. We will pose that the unification was realized by an informational substrate of the Universe, under the quantum level, the Universe being then a self-organized emergence on this substrate.

In order to define a unifying generation model of the Universe, we must first define a very singular initial state which engaged the creation of all space and all elements according to a generative hyper-process, then specify how the generative processes unfolded in a continuous and self-organized way. We posit that this singular initial state generator of the Universe is a localized informational system, that it was defined elsewhere and that it produced a quantity of informational elements which also produced some of the same. It was thus a generating system producing elements which in turn generated. We will then develop an informational model where all the structure and all the elements of the Universe will be considered based on informational fields of communication, allowing these fields to generate structures with temporal permanence and forming both informational and material elements. Furthermore the notion of information used will be different from the one defined by the Turing machines that run programs. We will use the notion of organizational information which will in turn use the notion of informational fields of communications, of the type, for example of a photon field but more intense and totally invisible due to operating at a speed much higher than the speed of light, and which will produce a space of dense communicational links everywhere between all the created elements. This space of informational links will be the substrate producing the space of the Universe. Therefore, there is an informational substrate that will perform incentive control at multiple scales, to continuously produce an emergence that will be the space-time with all the physical elements of the Universe being generated. It will be necessary to introduce a fundamental organizational law that will regulate all the formations and aggregations of generated elements, in order not to have a system randomly generating a set that becomes chaotic with very little structural coherences.

We assume that the generating element is an autonomous and complex informational element which will generate many informational elements which will be very communicative among themselves. This generating element will use informational energy that will be given to it by its builders and which it will diffuse to the elements that it will generate. The informational energy will therefore be the force that will activate all the informational fields and all the generated informational elements. The elements produced by the generating element will be of two types as will the production of these elements: either they will be structural elements and they will then form cells of the empty space which will constitute themselves on the informational links of the communications and which will unify to generate the space of the Universe, or they will be elements of activity, endowed with informational energy and whose actions will be to communicate to form aggregates constituting the material elements of the Universe, and they will duplicate themselves by producing other elements of activity or structure to realize the expansion of the Universe.

The generated elements of activity will contain fermions and will produce, by their informational communications, the quantum-type particles. So, from the initial generating element, there is the beginning of the production of space in the form of empty cells and elements of activity that will constantly produce others that will be elements of activity or structure, thus continuing the generation of space and quantum elements. All the elements of activity, which are of the type of informational fields, will be able to communicate to generate aggregates, that is, to produce atoms, molecules then all physical elements of the Universe. Thus the Universe generates itself, without ceasing its process of organizational deployment produced by the incentive of the organizational substrate which is its incentive controller at the informational level, because it imposes the application of an organizational law allowing the generation of structured aggregations.

We therefore propose a model considering the whole universe as a fundamentally informational and self-organizing system, composed of what we will call informational fields making up the elements of activity in the generated space, the Universe being in constant expansion. The structure of any produced element of activity will thus be considered as both material and informational, and the organizing informational fields will exist at all scales of physical elements. The Universe will thus be an organizational emergence on a substrate of informational energy that produces incentive control, which will be the application of the organizational law that we will define precisely.

In the model describing the generation of the Universe and all its elements, including the living organisms on Earth, by explaining the communicative action of genes, we therefore pose a central hypothesis which specifies that the Universe has an informational substratum, that everything is based on the generation and action of informational fields which use a basic informational energy and that this activity is subject to a law of multiscale incentive organization. We will show that the formation of the Universe can be represented by an informational program which generates its elements and uses a considerable dynamic memory for the control of the operations and which will be its informational substratum.

We will develop the reasons for the creation and evolution of life on Earth. The model will be based on the notion of communication between all the generating and generated elements enveloped by what we will call physical and informational membranes and which exchange information with incentive values specifying morphological modifications in these fields and thus in the generated organs. We will show that the organizational law can produce specific local bifurcations and that it is the case on the planet Earth which produced all living beings. We will show how the control that allows the modification of the organizations is carried out by introducing the notion of a morphological pattern, an informational element that alters the communications, that allows for modifying the action of the genomes and thus for evolution to occur, and that allows the psychic systems of human beings to have fundamental drives and tendencies. This will make it possible to show how and why terrestrial life was constituted, how and why sexuality was formed for the reproduction of organisms and how and why innumerable very sophisticated and intercommunicating species were formed, making up the Gaia system. We can then say that all life on Earth is based on an informational organization and that it is desirable that we apprehend well this type of information in its domains, in order for humans to position themselves in evolutionary social structures that are really shared by the use of fundamental communications, moving towards ethical, cultural and peaceful societies.

We will therefore present what this generating information is, how the informational fields operate in communications, how the informational envelopes of quantum and molecular elements are formed. We will present the informational law that allows the substrate of informational energy to incite the realization of material aggregates, then of stars and planets. We will see that we can use the notion of agent, which has been deeply developed in computer science, by defining informational agents representing physical elements in the Universe. Then in the second part, we will present how the generation of life on Earth was achieved by explaining why and how there was a continuous evolution of the formation of all species, how reproduction permits the generation of new organisms forming groups and then new species.

The informational model presented is an attempt to unify many scientific fields analyzing all the elements that make up the Universe, starting from the quantum elements and going all the way to the living organisms on Earth.

Part 1

Informational Generation of the Universe

1

The Computable Model, Computer Science and Physical Concepts

We will first specify the foundations of computer science considered as the science of the calculable, and then expose the general physical theories on the situation of the elements of the Universe.

Computer science, as a science, is based on the computable model of functions and compositions of functions, which is the Turing model. In its applicative aspects, computer science today has considerable technological applications that invest all types of production in the world, that have upset the use of communications and the manipulation and processing of the knowledge used. We will present the fundamental model on which the calculable functions are based and we will see that we must go towards another model of information manipulation to conceive at the informational level the generation of the space of the Universe and the elements constituting it.

1.1. The Turning model

Mathematicians and computer scientists have been interested in the classes of functions that can be calculated with algorithms, which are automatic calculation processes understood as sequences of instructions defining the values that the variables of the activated functions take. An algorithm is therefore a sequence of instructions that calculates the value of various specific functions, and is defined by its various steps.

The mathematician Alan Turing, in 1936, before the invention of the first computers, posited the existence of an abstract machine capable of calculating all the values of any mathematical function defined on the integers according to an automatic process. Such a machine consists of an infinite tape for storing data and has two parts – one used to read the new data and the other to store them. These data are numbers, which are interpreted by a reading system and written by a writing system and which transmits the read values to the instructions of the machine which uses them to produce the result, which is another sequence of numbers placed in the memorizing part of the tape (see Figure 1.1). There is thus a reading–writing head which makes it possible to specify the actions for processing the instructions. The machine is, at each step of calculation, in a state which is represented by a certain numerically indexed symbol and it is given a precise quantity of these states during its construction. The program of the machine is a set of instructions processing the read values to produce a numerical result. Each instruction has two parts: its trigger to activate and its action to process the value read from the tape. The correct instruction is activated by the trigger and it reads and processes the digital data that is being accessed on the read tape. Its action is to use and rewrite this value read in the same cell of the tape so that it can be used by other machines and then to move the reading head by one cell or not to move it, and then possibly to change its state to specify another one.

An elementary instruction of the Turing machine thus has the form of the following quadruplet (q_i, S_j, S_k, q_s) with:

q_i is the current state of the machine;

S_j is the piece of data which is read on the reading head;

S_k is the numeric character that will replace S_j;

q_s is the new current state of the machine after the replacement.

However, the machine can also have one of the following two forms, with D and G being the actions of simply moving the read head to the right or left without writing anything on the read–write tape:

(q_i, S_j, D, q_s)

(q_i, S_j, G, q_s)

This machine is totally automatic, and it is the most elementary possible with regard to the calculations to be carried out. It is the most important universal machine in the history of computation. All the algorithms use sequences of instructions and are therefore sophisticated compositions of Turing machines, and the instructions can lead to the request after their execution to place themselves on another instruction to be executed by the famous "Go to" and this repeatedly until meeting the instruction "End" of the end of execution of the instructions of the algorithm.

The functions that the Turing machine computes are called recursive primitive functions and they operate on sequences of natural numbers. They are obtained from basic functions, like identity, projection, successor function, using composition, recurrence and minimization, and they are executed in associations. They define all the usual arithmetic functions by machine associations, like power functions, products and sorts, and they are thus the basic model of what can be defined in mathematics to operate on sequences of integers.

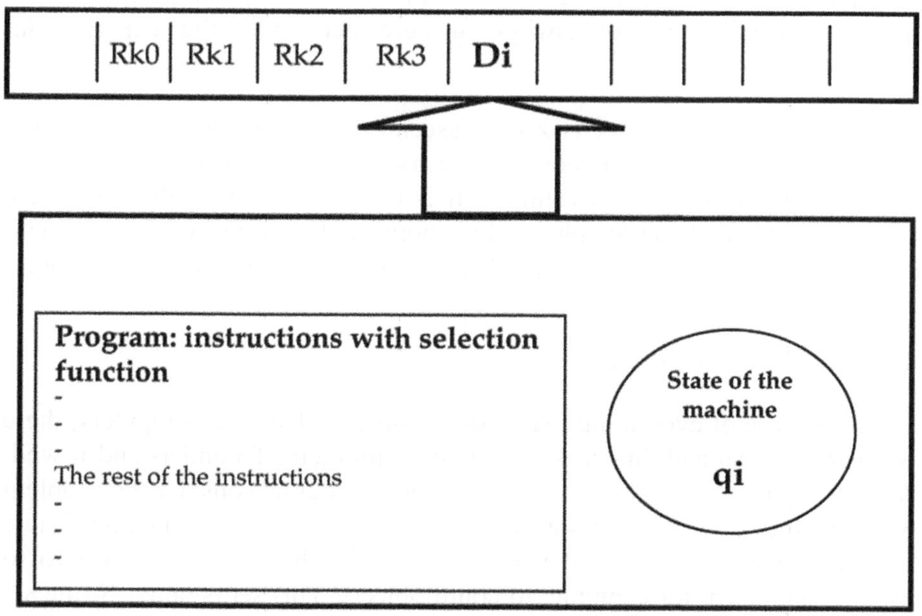

Figure 1.1. *The Turing machine*

All the calculations of the values of numerical functions are thus based on the design of Turing machines. We define a general function on a given problem and its effective calculation amounts to defining the set of Turing machines which will represent it.

However, we can proceed in a much more general way by using the differential equations of mathematics. In this framework, we first define functions and constants specifying the elements in relation in a physical system describing a natural phenomenon and we place these functions in differential equations representing the spatial–temporal relations between the observable elements of the phenomenon. We then seek the solutions of these differential equations giving the values of the functions and thus giving the solution of the problem of the relations and movements by comparing with the results of the physical observation to validate the equations. However, this problem does not always offer a good solution in fundamental physics which uses differential equations and partial differential equations representing the relations between the functions which are the characters of the studied problem, because the solution of these equations, if they are indeed calculable, is not always in agreement with the experimental measurements.

This approach is characterized as ascending, because we start from the observation of the phenomenon and we try to represent it by variables and functions that describe its evolution. It is therefore assumed that there is a space available and that the physical phenomena that occur in this space with structured elements must be precisely described to measure their evolution.

1.2. Computer science

At the base of everything that is done and calculated by computers, there is a very general and far-reaching scientific problem. To understand it well, we must approach the science of the calculable. Let us consider the problem of the mathematical calculation of integer functions. We are interested here in what can be done with the integers, that is, those going from zero to infinity, and used, for example, to count objects. This set is noted **N**, the set of natural numbers. We know, in mathematics, that we can calculate many things using integers, and we can define a lot of functions of the set of integers in itself. Indeed, we can code everything that is symbolic, everything that is cognitive with natural integers. Let us recall that any integer can be represented, if necessary, in base two, that is, with the two digits 0 and 1. We can generalize this and be interested in functions whose argument is formed by a sequence of n integers, the value of the function being another sequence of integers. In this case, we are interested in vectors of integers. We have thus defined all the functions with n integer arguments, and whose values are certain sequences of integers. We are in the domain of integer mathematics where any formula is in the form of a sequence of signs. We can encode this sequence of signs by integers and thus represent any mathematical formula by a sequence of vectors of integers. Any mathematical demonstration is, in the same way, a sequence of signs that can also be coded by integers. We can therefore see that the study of integer functions is a fundamental problem of the representation of mathematical language. The question is the following: since the formulas and the mathematical demonstrations of the domain of integers are finite sequences of signs, can we represent the demonstrations by programs? The answer will be yes, and for a very large set of functions and demonstrations.

Computer science as a science of the calculable appears here. All these integer functions, all that mathematicians can define on these integers in the form of various equations, are equivalent to computer programs of abstract machines. It has been shown that for any function of a sequence of integers in another sequence of integers to make mathematical sense, to be coherent, there must exist some abstract machine, an "abstract computer", with instructions that allow it to be calculated. The existential of all mathematical functions on integers has a meaning if the computable allows it to have one and vice versa. This very powerful theoretical result is the famous thesis of Alonzo Church, dating from 1936. It amounts to saying that for a function on integers to have a mathematical meaning, to be coherent, we need only define the program of a theoretical machine which can calculate it. If there is no such program, the function does not exist and is not logically admissible. Today, the fields of application of computer science are considerable, making it possible to represent practically all structurable knowledge in all fields and to direct quantities of electronic devices in real time.

In the usual approach, computer science deals with the processing of information related to multiple calculations, including those established from a great number of functions, with systems using as a basic element what is called state machines. A state machine is an abstract machine passing through strictly determined states, choosing them one after the other from a set defined as available, and in which precise elementary instructions are executed. We use the automaton by starting from an initial state to reach a final state which is the expected result of the following calculations. This notion of state automaton is fully used when dealing with problems that are decomposed into many sub-problems, all very well defined, the whole forming a perfectly structured set, where what is to be computed at each step is well specified and develops the continuation of what is to be computed. Such problems belong to the class of what are called **well decomposable problems**. Computer science deals, at the electronic level, with binary information encoding elementary instructions, forming the programs themselves composed of sequences of calculations carried out by elementary instructions. The length of the programs and the number of instructions of a program can be considerable, and several programs can easily be executed at the same time and communicate to each other at the right time information allowing to realize well synchronized calculations. But, in a classical way, any program remains a sequence of calculations which, step by step, each calculation step after each calculation step, passes through a finite number of predefined states until its final state.

This locally mechanistic vision of the computing process has evolved a lot. Today, we know how to make many, many programs communicate, based on state machines, which run in parallel and, above all, which modify their own machines during their operation by communicating to synchronize themselves, even though the basis of each program is still the state machine. We have therefore shifted the framework of the regular automation of programs to the notion of autonomy. We know how to build programs made of many sub-programs which have their own behavior, which can communicate, synchronize, modify themselves and which can especially generate new programs breaking the order given by the initially conceived state machines. These systems are called **adaptive multiprocess** systems, and they are the ones that run on current computer clusters. Indeed, this is the case for any networked operating system that manages the simultaneously active resources and applications of a desktop computer, which is so common today. The notion of multiprocessing is important, and it has been a basic notion for the conception of artificial consciousness (Cardon 2018), because it places the consideration of programs at the level of autonomous software entities, active, carrying out precise local actions and above all highly communicative with each other in order to form dynamic structures that are constantly changing.

We will use the following two different notions of the term process:

– The notion of a **functional process** which is seen as a vast movement of components exchanging information and energy, and producing the state of a certain system, as is the case of brains producing representations. We can thus speak of the process of emergence of a form of thought about something focused from a trigger generating intention.

– The much more precise notion of **computer process**, which is a small program wrapped in utilities and processed in a computer system that handles quantities of them simultaneously. We will then speak of swarms of processes to designate very numerous computer processes running in competition, this notion of swarm of processes being then close to the other notion of functional process.

Generally speaking, there are two categories of programs in computer science:

– The category of programs where it is a question of calculating a given function which is precise, well defined in advance, of strictly developing the calculations of all the necessary steps, which amounts to the execution of a structured set of state machines.

– The category of autonomous programs composed of multiple swarms of processes that will run in parallel, that will capture internal and external information, that will confront each other at certain times to exchange information, that will modify themselves, generate others and thus create new processes, to finally produce a global result that will be the most adapted to the situation that has evolved from its initial phase.

The second category is, for example, the state of all Internet users' programs over a given period of time, when these programs themselves constantly consult and modify highly interactive Web sites. There are no permanent elements in this case and the problem cannot be based on an *a priori* decomposition into permanent elements.

The difference between the two categories of problems proposed lies in a fundamental point. There are programs in both cases, but in the second case, they will have to be constantly modified, rewritten and evolve, whereas in the first case, the software is used as it was conceived, with its initial capacities well-defined and non-variable during use. The second case must make one think of a certain form of autonomous, very abstract, artificial behavior.

1.3. Formation of the Universe in physical sciences

In mathematics there exists the domain of real numbers, which is a complete Archimedean body, whose use defines sophisticated and very powerful equations: the differential equations and the partial differential equations. This domain of real numbers using differential equations has allowed physicists to define very fine theories of the evolution of matter in the Universe. To define a differential equation, we first define the variables that characterize the observed system, and then we define the functions and their relations that should allow us to predict the evolution of the values of the variables, taking into account the values of certain constants. All this is put into a differential equation which must calculate the functions and thus obtain the characters of the evolution of a system starting from an initial state. This will be used in a very important way.

The physical sciences have been working for a very long time on a generation model for the Universe. The main model of the physical theory describing the creation of the Universe is the Big Bang model, posed by Alexandre Friedmann in 1925 and by Edwin Hubble in 1929, then very widely developed thereafter. This model posits the existence of a primordial nucleosynthesis, a very singular set of quantum elements with a considerable temperature that inhibits the propagation of photons that continuously interact with quantum particles. After a hundred seconds, the photons lose energy and the protons and neutrons are able to associate in a durable way to generate the first complex nuclei of the elements. From this initial set, the Universe developed by a considerable dilation while its temperature remained very high. The initial temperature of the Universe fell to reach 3000 degrees after 380,000 years, producing light by the emission of quantities of photons. This light of the cosmological background radiation (Peebles 1980) is observable today, when the temperature of the Universe is only 2.7 degrees above absolute zero. The light formed by the photons therefore propagated out of this initial set by constituting the cosmological background, it propagated in a space whose theory does not specify how it is formed or where it comes from (Klein 2010). The classical model posits that the generic elements of the galaxies were created by this very particular initial set and that the Universe was constituted by continuous dilation, its expansion.

The Big Bang theory thus posits that the Universe originated from a singular point 13.7 billion years ago, but there is no precise notion of the formation of space or time, only a notion of considerable energy at this initial moment in the form of photon fields and the activity of quantum elements. Physically observable matter appeared from this initial state at a very high temperature, with neutrons, protons and electrons in considerable numbers, the protons being assumed to be seven times more numerous than the neutrons. Then were formed, by combinations, nuclei of hydrogen and helium. This set of elements will not cease to expand very strongly in space, thus lowering its considerable temperature and allowing the activation of physical processes that will form the nuclei of many atoms.

All this is defined by differential equations with constants, variables and functions, and these are the equations that give us all the characteristics of this expanding universe. According to these equations formulating the notion of energy, the temperature of several billion degrees Kelvin drops in a few seconds to one billion degrees Kelvin. Then, protons and neutrons are able to associate to form heavy hydrogen nuclei by releasing photons. Some of these nuclei are able to attract an electron to form a hydrogen atom. Some hydrogen atoms are able to attract each a proton, a neutron and an electron from their neighborhood to form helium nuclei, the helium atom being four times heavier than the hydrogen atom. It is deduced in this theory that 99% of the mass of the Universe is formed during this process, being thus constituted mainly of hydrogen and helium.

It is well-observed that hydrogen is very present in the Universe and this leads us to the following result: a quarter of the mass of the Universe is formed of helium and three quarters of hydrogen atoms.

The scientific theory states that the matter in the Universe represents 5% compared to the empty space which is considerable, but it also states that there is dark matter which is six times more important than the visible matter and which allows the maintenance of the conformation of galaxies and their structures which otherwise would not cease to dilate and therefore could not exist. However, this dark matter has not been physically found or observed. The theory posits that there is 70% dark matter and 75% dark energy in the Universe, which are still undetectable with current observational equipment.

In our model, we will pose a hypothesis specifying what dark matter is precisely. It is posed that the variations of density of matter will allow, by undergoing gravitational force, to form in the galaxies structures made up of stars and planets.

This description, describing how to generate the structures of material elements of all sizes, gives a very important place to chance. There is the chance of communications between protons and neutrons forming the first atoms, there is the chance of collisions between these atoms generating photons and constituting other atoms, then the chance of movements of atoms aggregating and forming small masses. There is finally the chance of the meetings of these small masses to form bigger masses, very big masses aggregating the smallest by the force of gravity, attracting to them the small ones and thus forming the stars and the planets.

In this general theory, everything is based on the atomic elements that can combine according to the chance of formation of aggregates of atoms that will form structured elements having permanence, like those of the stars and planets, and that are positioned in the available empty space of the Universe. But there are theoretical contradictions between quantum physics and the theory of gravity, as these two approaches to reality do not manage to agree by modeling domains of the Universe that are at totally different scales defined by equations that are not compatible.

The theory of the generation of our universe states that it is expanding. The theory is based on the notion of energy and expansive dilation but does not give any explanation on what is the space where the particles are located and how this space is formed. The great fundamental question is then the following:

– What is space, where does it come from, how and why is it generated by allowing physical elements to take place in it?

We are then going to specify that all that exists in the Universe is in the form of ***particular informational fields*** with deployment in a space that is going to be the dynamic reification of the activity of these informational fields, fields propagating unceasingly innumerable communications to realize by confrontation of the emergences of aggregations constituting elements that are going to have structure and will continue the realization of the changes of structure of the elements to generate the complexity of the organization. We will thus pose that the Universe is an autonomous system based on a dynamic informational space treating in a continuous way the specific informational communications between its elements to make them exist and continue to exist by becoming complex. The space will thus be the major constituent of the Universe, which will lead us to conceive an informational system different from the functional information of the Turing machine.

2

The Informational Components and the Organizational Law of the Formation of Space and the Elements of the Universe

We will consider the Universe as a totally autonomous system of generation of space and physical elements with a considerable size. The Universe was created by a very specific generating element, available, which engaged it to produce in a continuous way space and organized elements having spatial and temporal stability. We will show that the Universe is an organized emergence with informational components whose role will be to aggregate in physical elements on an informational substratum which realizes the self-control incentive. We will show that this generation is done following an organizational law that is at the level of its informational structure. The informational model that we present will allow for considering the Universe as the continuous generation of a self-organizing system that creates space, matter and physical elements, based on specific informational communications that are absolutely continuous. We will show that we can also make a multi-agent informational approach to represent the generation of the Universe and all its elements, from atoms to life on the Earth.

2.1. Informational model of universe generation and organizational law

We will assume that the elements of the Universe system are of informational-type. Basic elements will produce space and generators of quantum particles, and then the aggregations of these quantum particles will produce molecules by a general specific control of their organizational actions. We will thus pose that the basic elements generate space, which will be realized by the production of elements forming informational links in the form of fields and which will also be spatial cells. The basic elements of the Universe produce therefore space which is informational, and then in this space will also generate elements of activity which will aggregate to produce atoms and physical elements. To produce all this, we will consider that there is an informational substratum in the physical universe and which formed it by emergence. There will be thus two fundamental parts in the Universe: there is the informational substrate which is everywhere under the space-time and there are all the physical elements which form a spatial emergence on this substrate.

We will first specify the two fundamental concepts of our model, specifying what is the foundation of the Universe in this modeling and what is the basic informational character.

> THE 1ST CONCEPTUAL PRINCIPLE. – The foundation of the Universe by a generating system
>
> There is a universe generation system, a creation realized in another universe with another space-time. This generating system has considerable basic informational energy and it will continuously produce an unfolding of local informational fields in the form of informational components and a substrate, which will be constituted by active informational links between all these components and which will realize the control incentive and which will constitute the new universe. These informational fields of the components have their own frequency and structure, which will allow the generation of space and innumerable physical elements by the emergence of aggregates of components. These innumerable informational fields of the components will fix the notion of time and space in our universe which will be created as an emergence. The Universe is thus generated according to a continuous informational construction incited by the substrate, it is a considerable emergence of aggregates having stability on a substrate of informational fields whose control will form the structures and produce the organization.

Let us specify at once that the notion of an **informational component** defines the central element of the informational model of construction of the Universe and that there are two quite different types. There will be the **activity component** which will be the generator of the quantum elements and there will be the **structure component** which will be the generator of the space cells of the Universe. We use the term component for these two different entities, because they are both informational fields that will be constructive in their specific domains and they will always be generated, either initially by the generator component of the Universe or then by the so-called activity components which will be continuous generators.

It must be made clear that the central character of the Universe is strictly informational and that there is a substratum that allows its physical emergence.

> THE 2ND CONCEPTUAL PRINCIPLE. – The informational character of the Universe and its substrate
>
> The universe generation system has a basic element, in its universe, which will emit informational fields which will be the first informational components, which will be virtual elements for our space-time and which will also constitute the elements of the informational substrate of our universe. The substrate will be constituted by a network of active informational links communicating with all the components and thus all the generated physical elements which will be aggregates of components. This substrate will be a network of active informational links under the space of the Universe exerting a general morphological and incentive control. The whole of the informational fields will thus form very dense clouds and constitute an informational mesh everywhere under our space-time and constituting the informational substratum of the Universe. These informational fields have a frequency which is outside the notion of time of our universe, and they will be in the form of virtual informational components for us and will generate aggregates having structure with permanence. These informational aggregates will communicate with the substrate which will be able to transmit information and informational energy to them increasing their intensity, so that they are in action following an organizational law of creation of our

universe. All these clouds have a function: to represent the organizational law that generates the organization of the Universe.

The informational fields are thus the basis of the substrate, and they have a frequency that is not accessible in our space-time to found it as a distinct world. We will state later on another fundamental concept on the structure of the informational particles generating the hyperdense flows of information, of which the Turing machine is a different representation because it is essentially functional.

It is therefore necessary to define in the model which is the original element that generated the unfolding producing the Universe, how space and time are formed, how particles are formed and how particles organize themselves to form stars, planets and unfold life as on Earth. The explanation of all these processes will be developed in our informational model.

It is thus posed that the Universe was created by an initial generating system and that it is its emergent production of continuous activity endowed with very strong organizational autonomy, with a radical informational distinction between the generating system and its emergence forming our space-time universe and which does not allow the elements of this emergence to communicate directly with the informational elements of the substrate.

There is mainly a general **organizational law** coming from the informational substratum of the Universe and which will be applied to make the generation of the Universe self-organizing, which will engage by incentives all the informational elements localized with permanence to aggregate and make multiple material structures, like molecules, physical bodies, stars and planets where life will be able to unfold by being the continuation of the unfolding of this organization which is realized at all scales. The Universe that we observe is thus an emergence realized on a substrate and the whole of the informational elements. Let us note well that this substrate will have the fundamental role of launching information of incentive control to enforce application of the organizational law and to carry out the organization of all the components in aggregates and in innumerable aggregates of aggregates. We will develop this point in the following by presenting an agent model of the Universe system.

THE 3RD CONCEPTUAL PRINCIPLE. – The organizational law of the generation of the Universe

Considering the Universe as being formed by innumerable elements whose foundation is informational, which communicate with each other to aggregate and generate new elements, including elements of great size, it is necessary to posit that there is a general principle of deployment of this universe that incites it to form constructive and well-structured aggregations, to carry out and amplify certain developments and to reduce others, so that the whole is, locally and globally, in good organization and is not chaotic. This principle will be represented by an organizational law that will be exercised by the informational action of control of the substrate and will impose that the components are in continuous organizational communication forming aggregates and realizing their structural unification. The control will be morphological and will be a general tridimensional regulation of the constitution of the Universe. We will call this incentive control that makes the organized generation of the Universe an organizational law, and we pose that there is an informational substratum that makes this law be exercised everywhere under the space-time of the physical reality to realize the incentive control.

The Universe is thus generated following a law that enables its coherent and organized deployment with good structures that tend to have temporal and spatial stability, and permits an emergence of physical elements possessing stability on an informational substratum that realizes its incentive control. Let us note that the control is absolutely not directive, as is the case in the majority of the functional systems of our societies, but that it is simply incentive and is exerted morphologically, produced by continuous information coming from the substratum which engages its particulate and structural realization.

We therefore will define an organized system with self-control given by an organizational law that is exercised everywhere and allows the Universe to be constantly conformed at all its scales. It is a system that develops and organizes itself in a continuous way, starting from the disposition of an initial generating system possessing a considerable informational energy and that will produce informational elements that will multiply themselves by propagating information and informational energy and by generating space-time. It will thus be necessary to define well what are these elements which generate space and the elements generating the fundamental quantum elements allowing the generation of material elements. It will be necessary to redefine a model of informational system different from the model of the Turing machine by presenting another approach of the functional duality between a computing center and the data ribbons, by showing that what will be the correspondents of the data ribbons will be elements of fundamental informational computations which will make the organizational law exist.

We assume that there is an initial informational component that generates the whole universe, which will generate innumerable other components that will all have an essentially informational structure. Let us specify what is the informational component, which is the basic element of the whole universe system apprehended in the informational sense.

Notion of the informational component of activity and structure and the emergence of the Universe

An **informational component is** a strictly local informational element that has been generated by another informational activity component and uses the energy and information given to it. There will be two types of informational components, the **activity components** that will produce the quantum elements, and the **structure components** that will ground the space of the Universe and envelop the substrate. The activity component has informational energy, and it sends elementary information and will generate other components. Its generating activity will produce either components of structure forming space cells, or components of activity which will be the basic generating elements of atomic elements. The informational component is the element that will realize the **emergence of the elements** in the form of space or quantum particles, they will be the basic elements of constitution of all the elements of the Universe. In its generations, it will unfold the substrate formed by the informational links between all the created components.

The whole universe will be an unfolding on a substrate that will be the provider of what we call **informational energy**, which is the fundamental force allowing the construction of the Universe.

> THE 4TH CONCEPTUAL PRINCIPLE. – The informational energy of the substrate of the Universe
>
> The initial system that generates the Universe must continuously send to its generated informational components what we call informational energy. Informational energy is a local informational impulse, more or less strong, containing precise information, imposing on the receiver a commitment to a certain action by giving a simple message. This energy is available in the informational links forming the substrate of the Universe and they are active links to make the organizational law apply everywhere, the substrate being under all the components of activity or structure. The informational energy thus forms a set of intense flows making all the informational components of activity communicate between themselves, by generation and transmission of information to the other components to generate organization. This informational energy will be the character of activity of the dense informational layer of the substrate of our universe, under the space-time and under all the physical elements, by allowing them to produce by themselves photonic, quantum, electromagnetic energy. On this substrate of informational links and informational energy, the Universe is indeed an emergence of physical elements of activity which have autonomy, with production of innumerable elements endowed with physical energy for their movements and their transformations in the space-time. This informational energy has been produced by the system of conception of the Universe that has provided it to the generating element, that has transmitted it to all the elements that it has generated and that has been then distributed in all the generations of the components. It is the informational energy of what is called dark matter which is the substratum of the Universe.

The substratum of the Universe is a set of active links that realizes the informational unification of all the components and that realizes the incentive control of their aggregative actions by the organizational law to realize continuous self-organization. It is therefore a system that is totally the opposite of functional control systems where a central entity analyzes all the elements and makes them activate according to certain commands. The substrate is an underlying control system, totally distributed and with a simple incentive, which is under all the components of the Universe, linking them by informational unification so that they activate themselves in an autonomous and organized way, each one of them thus having a relative autonomy.

We can then specify how this energy is given to the informational components.

The informational energy given to informational components

Informational energy is a communicative flow available on the informational wire fields forming the substrate of the Universe actively linking all components. The wire field of the substrate surrounds each active informational component so that it is activated by this provision of communicative informational energy, generating messenger informational fields with meaning sent to the other components. Each flow of informational energy sent to the components by the substrate is in the form of active morphological links that activate them, containing fundamental information that makes them activate in an organized way according to the state of their context that the links of the substrate can grasp. This is the key to morphological control.

The substratum is thus the basic informational system that continuously produces the active informational wire fields that create the informational components that are informational reifications, by engaging them to aggregate to produce all the organization of the components that form the Universe. We can then specify the difference, at the level of the concept of produced information and of used and produced data, between the Turing model and the model of informational generation of the Universe.

In a Turing machine, an information is a simple piece of data sent on the input or output tape by the calculations made by the internal program of the machine. In the substrate of the Universe, there is a complete mesh of informational links connecting all the components and these links are active by containing information that they have created and that makes the organizational law apply continuously. These are therefore active informational action links and not simply transmission links, and hence, we assume the existence of informational energy in order to distinguish informational data from its generation. The components will also be able to generate information to send it to other components by using the communication links. In the informational system of the Universe, there is thus mainly the generation of information, and the transmitted data are transitory flows. There is no structural distinction between a Turing ribbon and a computing center because everything is computational and produces information in the Universe system.

We can therefore say that the Universe is essentially calculatory at the informational level, that it produces everywhere and unceasingly local calculations so that elements may be generated and so that these elements may be organized and constitute stable forms. It is conceived as a calculator operating on itself and for itself, and it is rather an **informational organism** than a calculator.

It will be necessary to show how space is formed and constructed, how time is constituted and how the basic quantum elements endowed with informational energy fields are formed, and then the aggregates of these elements which will constitute the matter forming all the organized things of the world. It is therefore necessary to define a general unifying theory, taking into account the management of all this and not limiting ourselves to the notion of production by chance in a space that happens to be there, totally neutral and available. It will be necessary to specify how the elements can be activated and we will pose that they have informational energy which will be provided by the informational substratum that lies dense everywhere in the Universe. We will thus pose that the physical universe is an emergence organized on a systematically dense informational substratum.

2.2. The notion of generating information in the Universe

The model we propose is based on the notion of continuous processing of fundamental information between all elements that are continuous producers and readers of information and this information is essentially represented by local information fields.

We posit that there will be under the physical universe a substrate of informational energy unfolding during all its construction which will create all the components. We can then state the fifth fundamental concept of the informational model, which poses the existence of the generating element of the Universe, and define its role:

> THE 5TH CONCEPTUAL PRINCIPLE. – The generative informational element in the Universe
>
> The Universe generation system is based on the disposition of a generating element that will produce the first informational components, the beginning of the substrate and the space. This element of creation of our universe comes from another universe and is launched as action to create a considerable informational emergence which will be our universe. This generating element and all the informational components both generated and themselves generators have the inverse structure of a Turing machine, they generate a set of communication links in a continuous way between the produced informational components of activity, forming thus the informational substratum that will unite them and allow the realization of the substratum and the space. For these generated activity components, their action is represented by an informational envelope that is local to them and they are located in spatial cells. The generating component will thus generate a lot of activity components and structure components, the activity components will multiply themselves and the structure components will envelop, by the space cells that they form, the generated informational links constituting the informational substrate of distributed control of the whole universe.

We will thus have to specify how our space-time universe was formed from this generating component. The Universe was created by the generation of a considerable set of informational components of activity and structure produced initially by the generating component, then the components of activity which will multiply will generate the physical elements by producing aggregations and each generation will also produce informational links deploying the substrate. We will specify all the characters of the model creating space, quantum elements, and then the great physical forms by using the informational elements of the substrate.

Let us specify what the informational field of a component is.

The information field of a component

An informational field is an active dynamic form representing a basic piece of information and it is transmitted by informational links of the substrate. This field carries generating information whose character is given by the morphological conformation of the informational field of the receiving component which interprets it and can follow the character proposed in the information. This dynamic field has the form of a quantum field, but it is considered at the strictly informational level. It is thus the propagation of a form with strictly morphological characters that give it its meaning. The morphological form and the magnitude of the informational field thus represent its meaning and its importance and any deformation modifies its meaning, which is in all cases elementary. Any emitted field is readable by the structure of the field of another receiving component by its own field, any component having a set of structured local fields that modify and conform according to the form coming from the emitter thus realizing its interpretation. The reception of any informational field is therefore the constructive deformation of the receiver's field.

The Universe that we apprehend and of which we can physically measure many elements is thus an emergence of material elements having stable forms on a considerable network of informational fields, which are multitudes of elementary fields. These fields are not directly accessible physically by an atomic action that can only use them and modify the behavior of an observed atom. They are absolutely elementary and can be considered as subatomic.

We can specify what the informational substratum is on which the Universe is based and which will continuously generate active informational fields.

The informational substratum of the Universe

The informational substratum that engages in the creation of the physical universe is a network of informational links carrying informational energy that will produce an emergence of elements that will form the Universe constituted of physical elements having a certain permanence. This substrate is a considerable network of active informational fields which will link all the informational components of activity, these very specific fields not being directly perceptible from the observation of quantum or molecular elements. We must consider that from the initial element generator of the Universe, which possesses a considerable informational energy, there is generation of space cells which will be spatial components and new basic components of activity to which it transmits informational energy with the creation of local fields reaching them, which will allow them to activate and multiply. The initial element thus generates components and the informational links of the substrate connecting them. All these generated components are connected by the network of informational links carrying informational energy and information and constituting the production of our **universe substrate**. The generation of a physical element will lead to the creation of an active informational link between its generators and itself. Thus, any generated element is linked to all the others already created by the network of informational links of the substrate. The Universe is based on a substrate that has the structure of a complete hypergraph of informational links between all its components and elements, allowing them to be made active through the continuous propagation of informational energy and indications of organizational activities. This complete graph is the **informational substrate of our physical space-time universe**. We specify that the substrate, by its systematically dense network, will send to all the components information of the control-type in order for them to apply the organizational law: the Universe is a self-controlled autonomous system produced by a system of generation based on the initial action of the generating element. We can specify that **this substrate is the dark matter** considered in physics.

There is indeed an informational substrate holding the informational energy of the Universe, and this substrate is the dark matter considered by physicists.

We will develop the action of this substrate in the following by analyzing how the generated components behave. Let us note that when the space-time of the Universe disappears, which is the case in certain explosions of stars where the quantum, molecular and material elements as well as the informational components disappear, we obtain what is called a **black hole**. In these vacuum cells, there is only informational substrate, informational links that are not active because there are no active informational elements to use them. There is no more space nor time nor any physical or informational element in these cells and it is impossible to observe the interior with physical elements since there is no more space-time inside. It would be necessary to introduce informational energy and active informational components to reconstitute things.

We thus pose that the Universe is founded on the generation, the exchange and the treatment of basic information in a system which generates and manages the sending and the treatments of this information to carry out in a continuous way its development and its organization. In the Universe, all these exchanges of generating information will be in the form of informational fields. The basic elements that will emit construction information will be the more complex components formed by structured sets of information fields that will be located to constitute the elementary quantum elements of the Universe. These components are thus the basic entities generating the constituents of the quantum particles, whose structure we will specify finely. Let us give the general properties of the informational component of activity.

The activity information component and its information envelope

An informational component of activity is an element having permanence and which is constituted of some local elementary informational fields forming its structure. These elementary fields allow it to carry out its current action and characterize its type which will specify which quantum element it will be able to generate. All these fields which form its structure allowing its specific behavior are active in what we call its **informational envelope**, which is the specific field unifying its internal fields and which is deployed in the informational space around it with the structural components representing the space and the control elements of the substrate. This informational envelope allows its existence and ensures its communication with the other components via their envelopes and informational links so that aggregates are formed that will constitute particles. If the component is active, its action is to duplicate itself by creating similar informational components of activity and structure and informational links connecting them all and to communicate to incite the production of aggregates in its context. If the component is of structure, it is simply a constitutive element of the informational substrate and of the spatial void space carrying the fundamental information of the organizational law.

This basic element is absolutely tiny and it is virtual for the physical measurements made in our world, it has the minimal dimension for all the dimensions of the elements of the Universe. This component is either of activity to communicate and associate with others, or of structure, being space vacuum available to carry the information of the substrate.

The two categories of information components

1) The informational component can be an **activity component** in action and this action produces the generation of new activity and structure

components, and realizes communications with the other activity components to generate aggregates that will constitute the basic physical elements of the Universe.

2) The informational component can be generated as a **structural component**. It is then an informational component whose role is essentially spatial, it forms a space cell. It is in the form of a space cell valid for informational links which are fields and its size can be much larger than that of the activity component. It will be in informational union with the other structure components to allow the propagation of the information of the fundamental rule of the organizational law via the informational links and for the morphological control of the activity components. It will thus induce the sets of activity components to form aggregates according to the morphological considerations of the spatial domains where they are or rather to produce structure components to develop new spatial void domains. This component allows for conveying the organizational control information by uniting with all the other structure components and by forming the space of the Universe.

The organizational rule in this state of the beginning of the Universe is thus to generate vacuum deploying the network of communication and control of the substrate and to multiply the components of activity and to aggregate very many of them so that the quantum elements are formed.

> THE 6TH CONCEPTUAL PRINCIPLE. – Self-control in the organization of the Universe
>
> A generated informational component of structure is merely spatial and is then considered reduced in the sense of quantum creation. However, this constructive inactivity does not reduce it to inertia, because it is a basic informational element of the formation of the general informational network of the substrate which will convey information and organizational energy to engage the components of activity to aggregate, to transform themselves, to organize themselves by sending them messages, after analysis of the morphology of their groups. This component, by the use of the substrate, will transmit informational energy and the information of the organizational law to the activity components. We must consider that these structural components forming the space of

> the Universe are in fact the links of informational action of the system of self-control which apprehends the morphology of the groups of components of activity to make them organize themselves. There is indeed a continuous self-control exercised by the space which envelops all the components of activity and all the physical elements of the Universe.

We are thus in a space which constantly conveys the information of morphological organizational control and which forms the informational substrate of the Universe under the physical space-time. The space-time universe is an emergence of physical elements organized on a hyperdense substrate of informational control space. Let us insist on the fact that there is absolutely no centralized control coming from the Universe creation system, there is total autonomy and self-functioning from the creative action of the generating component and the generated components.

The activity of an informational component manipulates what we call generative information to contact another component.

The information that generates

The basic information used by the components and carried on the informational links is called **generating information**. It is **an action signal in the form of an informational field** generated by an informational component that contacts another and makes a link with it that will eventually have permanence in order to constitute a particle. Let us recall that all the generated components are in connection and that any component can solicit any other. The generating information is not a signal in the form of data but it is an active morphological informational field, the form of which will exist between the two elements as long as they are active and will not send new information to each other, the two elements becoming thus linked by the signal sent by the transmitter when the receiver accepts the action of linkage by following its meaning. This is the memorization of the information sent and received by the components; it is a memory of the Universe's information system.

This notion of used information is thus different from the information proposed by the Turing model, because it is not at all in the form of a sequence of binary punctual elements. It is an active information field whose morphology is variable according to its meaning and which is processed by the information fields of the receivers which are deformed to carry out the processing.

In our model, the generating information is conveyed by a morphological link which exists or is realized between the elements and it is used or not according to whether the receiver accepts the reception of the basic information and associates with the transmitter. By that, the notion of hyper-complete graph memorizing the exchanges of information between the whole of the informational components of activity is constituted, and it is thus the dynamic memorial structure of the substrate of the Universe.

The central concept of the model that we propose is to consider that the Universe is a system with a totally informational base, that it is constituted by innumerable local informational fields constituting the envelopment of the generated informational components. The Universe unfolds and the notion of space is of the continuous creation of informational extent constituted by basic components of structure, which is realized so that the precise fields where components of activity will realize the material elements may be constituted.

The space-forming structural component

An informational component of activity has an action that allows it to generate others, which can be created as being of activity or structure. A structure component is an element that is a vacuum cell and that does not act by creating components but that allows the general control action to be carried out between all the activity components. It supervises the informational links of the substrate in its spatial domain. This structure component is linked in an informational way to its generator, and thus to all the other created components, and it is thus well connected to all the structure components that form space. The connected set of all these structural components that carry informational energy through the substrate will constitute the **spatial vacuum of the Universe** and will allow incentive control to be exercised on the action of the activity components, by applying the organizational law. The **incentive control** is thus realized by the sets of components of structure enveloping spatially the components of activity which, according to the size of the field of these components of activity, will incite them by the sending of a basic information to generate aggregates. This is indeed an organizational control performed by one set of components on another. These components of structure conform spatially with the informational substrate of the Universe allowing its self-organization, by inciting the components of activity to generate either components of activity and to make them aggregate, or components of structure to increase the size of the space by allowing thus the components of activity created to be in domains of very good spatiality.

The notion of generating information that we propose is different from the notion of information sent from a transmitter to a receiver in the form of neutral data by electronic devices and which is the usual definition of Shannon (Shannon 1948). In the calculable, we assume that systems manipulate and send information in digital form generated by devices containing computer programs and using well accessible memories, using multiple algorithms containing instructions, which start from an initial state to arrive at a final state. In our model, we will not consider machines that have a space for reading-writing data available in memory structures, but we consider active informational elements that have their domain of informational values in the envelopes of the informational fields that constitute them and that are located in spatial domains constituted by components of structure that generate the information of control and transmit the information of the components of activity between them. For the basic physical components of the model generating the Universe, their foundation will be to create components of activity and of the informational space by localizing their fields there. They are thus informational generators which build their field of action, which is their space of transmission of information. Thus, the information will have a morphological and energetic form and will move in the spatial and informational substratum of the Universe; it will be the memorial flow of the informational activities of the basic components in action. Informational components of the system will gather in the form of structures unifying their fields to constitute the quantum elements, which will be the first stage of element generation. Then, they will continue their actions by following incentive rules to constitute and generate all the material elements at all scales. This notion of generating information is the following.

The generating information creating the Universe

The Universe system is a system that generates and processes a certain form of dynamic information and it is the generation and use of this information that makes it exist. This information is in the form of active morphological fields whose forms represent their meanings. All information is temporarily reified in the form of field lines that link the components forming a considerable deployment. There will be generating components whose action will generate structural components forming space, and activity components which will organize themselves by their communications and form the elements of physical matter by structuring themselves in the available space. There will thus be informational energy available for the action of all the elements which will be given by the informational substrate of the Universe linking all the components by forming a dense web everywhere. In the Universe system, a communication information between components is a local field generated by the action of an informational component, it is a field which represents, by its morphology and its amplitude, elementary characters of organizational activities to be realized. It is thus a ***constructor informational field***: this generating information is what organizes everywhere and at all the levels the physical elements of the Universe, which must thus be regarded as a very organized whole of generating information produced and interpreted unceasingly. This constructor field uses a weak informational energy to generate new elements and to link them together by the generated informational network.

We can then specify the notion of organizational tendency that makes the Universe be.

Organizational trend of the Universe and informational envelope

The basic components that carry out the generation of space and all the elements of the Universe follow a tendency to the organization that is the major rule. It is a question of realizing and manipulating sets of elements constituting informational forms which have structures presenting coherence, by memorizing their structural form in an informational envelope. Therefore, there is the fundamental notion **of informational envelope**, which is a union of informational fields enveloping an element that has existence and stability. The extent of the informational envelope of an element takes space and has an energetic and memorial character within the Universe. The envelopes will be realized at all scales, realizing the union of the basic elements, then the elements constituting the atoms, molecules and material elements, and finally the stars and planets orbiting around them. The basic elements of the substratum will send information to the components to carry out actions in accordance with the organizational law; this will be the control of the organizational tendency of the Universe. We will see that the key to the existence of all life on Earth is the systematic memorization of the plans of conception of the generated organisms to allow the continuous evolution of these living organisms from the evolution of multiple plans of conception. The Universe is thus an **organizational system which memorizes its informational structures created in the form of physical elements**, by memorizing the links of communication, to realize the continuous evolution of its organization.

Thus, the Universe is constituted by the action of the basic components of activity which will have two very precise actions.

Constructive informational actions of the activity components

1) **First category of action**. The components of activity generating the Universe are in the initial action of generation. This action amounts to generating an informational component of activity or a component of structure to form informational space. The component of activity will be able to duplicate itself by creating a component which is similar to it and which is active, and this in order for there to be a significant number of basic components of activity in the informational space and to be able to pass easily to another stage of activity. It will place this new component in the space that will be available that it will generate by creating structure components. In this generation, it will use either the informational energy of the substrate of its domain limited by its informational envelope or it will eventually receive some from the substrate.

2) **Second category of action**. Initial activity components in their informational envelopes in the available space created by the structure components communicate to associate and form the first **aggregates** of atomic elements, consisting of several basic components that structure themselves. The actions to achieve these associations are to constitute and then maintain the formed structure and then continue to seek to associate with other components by sending information through the informational links of the available space to form more complex structures, to form the molecules. They will always receive information from the substrate related to the organizational law to follow its direction, in order to organize the elements.

These two types of activities are the two characters of the actions of the informational components of the Universe according to the state of their domain.

2.3. The generative informational component and the informational energy of the substrate of the Universe

We have specified that there was initially the activation of a very specific generating element, which we have called the **informational generating component of the Universe**. This creative element is endowed with a very great informational energy and it will proceed to considerable generations by producing a set of informational components which will themselves duplicate and then associate and form the material aggregates. The generated active components will have complementary atomic types so that they can form atoms by aggregating. At this level of description of the model, we still call all the generating elements informational components before we move into the realm of informational agents. The basic generative element that produces the first informational components does not come from our space-time, it comes from another universe where it is designed and activated and is put into action with its very strong generation rules to allow the launching of the effective generation of the space-time, the substrate and all the elements necessary for the continuous creation of our universe. This is the basic assumption for the foundation of our universe.

The informational components are elements of informational activity made to multiply and communicate between them and realizing, by these communications, the structuring and the organization of the Universe. They are the elements that will carry out all the characters of the program of generation of the Universe by starting the global organizational process. In the initial stage of action, the generative informational component is activated and, with the considerable informational energy supplied to it, it will generate many activity components and structure components. The activity informational components will generate others by themselves and the structure components will produce the deployment of the space. The whole of these elements will reach a considerable number in a space which they will create unceasingly and make available, the components of activity will be able to communicate by creating thus swarms to carry out the structured generations of atomic then molecular elements. These components will communicate and associate to produce all the elements that we call the material elements of the Universe system.

We must therefore assume that the creators of the Universe have generated a very sophisticated element that has informational energy and manipulates information to produce informational elements that have permanence and are located in an empty space that it creates by supporting an informational substrate of control. We will come back to the structure and the realization of this generating element.

We posit that the initial stage of the creation of the Universe is the action of the generative informational component endowed with a considerable informational energy and which will generate new components which will be of activity or structure. Our hypothesis is thus that the Universe starts from a singular element endowed with very great functional and organizational properties to create three-dimensional space and components which will form the basic quantum elements.

The generative informational component and the informational substrate of the Universe

Initially, the designers of the Universe system must produce a substrate of informational energy with a specific frequency, which will become dense everywhere under the generated physical universe and which will contain, by its diffusion, all the informational energy necessary for the generation of the elements that will be produced. The designers have created the generative **informational component** by endowing it with a considerable source of informational energy, it will activate with a flow of energy to produce a considerable set of **informational components** that will form space and also generate the basic quantum components. The space, the union of the structural components, will be done on an informational substrate that will contain informational energy and will initiate communications to satisfy the organizational law. The created activity components will all have the property of applying the organizational law and will have the different basic quantum characters necessary to form the elements of the generated universe. All these informational components created as active are of the informational field-type, they are endowed with informational energy coming from the substrate under the structural components, and therefore, they have a structure to activate and be able to aggregate and form the first atoms. There will therefore be, in the multiple spatial zones created where there are many components of activity, the generation process of physical elements by aggregations of these informational components.

The informational energy with which informational components are endowed is a very specific energy:

Informational energy and the substrate of the Universe

At the time of their creation by the generating component, the informational components of structure are located in nothingness and constitute, by their generation, an initial spatial cell. These cells will be able to localize the informational components of activity. As any activity component is a set of informational fields, its capacity to communicate or its capacity to duplicate itself by generating a relatively similar element does not require energy in the physical sense of the term but simply requires the capacity to generate a new small informational field, a small sheet whose morphology gives information. Informational energy is the **fundamental informational force** given to any activity component to exist and activate. This energy is a set of informational fields transmitted by the substrate to the generated active elements. Informational energy is therefore the ability of any informational field of activity to produce an informational action. This informational energy allowing to send the right information at the right time to the right elements is given initially to the generating informational component, and then, by the substrate which transmits informational energy, to all the components which are generated. Therefore, there is a creative system under the Universe that gives informational energy to any generative component and then gives it to all the components that are created via their communicative network, and that allows the activities of all the components to be carried out by ensuring the exercise of the organizational law. This is the key to the generation of the Universe in continuous organization and expansion.

The space of our universe is created by the generating component which produces informational components of structure and activity, the components of activity will duplicate themselves by receiving from the substrate a fundamental information soliciting their action of organizational generation. There is thus the creation of the network of links between all the components of activity.

Informational links

During the creation of an informational component of structure, there is creation of a spatial cell. This spatial cell will represent a form of memorial capacity of the creation of a structural component by its generator, and which will be linked to other spatial cells. These spatial cells will form, by unification, the physical spatial space which will be a considerable network of informational links between all the components of activity, by being elements able to convey informational energy and to ensure the transmission of informational messages between the components. We must thus consider that the space of the Universe is based on a totally dense hypergraph of neutral and available informational links between the components of activity, usable as a network of information transmitters, which ensures the capacity of general informational action of the substrate on its emergence constituted by the physical elements forming our physical universe in the space-time. This network of available informational links thus forms the considerable dynamic memory usable for the actions of information transmission between the components of activity founding the Universe. This is the informational foundation of our universe on its communicational substratum which is dynamic.

We can thus define the characters of the physical emergence of the Universe on its informational substrate. We can consider that the Universe is a strongly organized emergence of the whole of its informational components of activity on a uniformly dense substrate formed with the components of structure and containing informational energy reified in the form of links between all the elements and which allows for all their activities to be carried out by impelling the multiscale organizational law. This **informational substrate** is the means for all the components in activity to take informational energy and to act. There will be systematic creation of informational components in the periphery of the global cover of the generated components of activity, by the sending by the substrate of a fundamental piece of information giving the incentive to the creation of space then of components of activity. We see thus that the Universe starts from its generating element and is in continuous expansion of space and elements by producing an emergence in a continuous way.

The substrate is therefore the basic energetic and informational system initially associated with the generating component and which will produce, without ceasing, informational energy to make all the informational components active. The substrate is not a classical space with predefined dimensions to represent distances, it is a hyper-general generator of informational energy that is available to all the active components of the Universe when they need it to perform their actions. The substrate of the Universe is the system that gives the informational energy to all the components and its property in the organization of the Universe will be to generate the photons carrying a minimal quantity of informational energy and which also give, by their limit speed of displacement in the vacuum, the key of time.

Organizational properties of the substrate of the Universe

The informational energy substrate of the Universe is an entity present everywhere under the space-time of the physical universe connected by the links between the components and which diffuses everywhere the informational energy necessary to the elements, in the form of available photons and of active informational fields carrying elementary information. This substrate will produce, at the beginning of the creation of the Universe, a considerable flow of photons which will activate the initial components so that they duplicate themselves and communicate by using their informational energy. This substrate will continuously give informational energy to all the components to produce the aggregations forming the quantum elements and then structured matter.

The relationship between the substrate of informational energy and the space-time of the Universe made up of components of activity or structure will be done by the direct transmission of energy and information to the informational components and by the generation of photons, which will be elements carrying informational energy. The **photon** is a very specific local field of informational energy, which is considered in physics to be both of zero mass and electromagnetic energy. It is produced by the action of the substrate of informational energy of the Universe in the action between informational components, being used as a specific transmitter of the greatest specific informational energy given in an action. It is generated by the action of these components, and then it is used by others. The photon has no notion of mass and moves in the vacuum at a speed that is constant, which is the possible limit speed for all active elements in the Universe. This speed of displacement of the photon will in fact give a major character to our **notion of time**, any notion of time being the duration of an event or of the displacement of a particle in relation to sources and which will always be lower than this limit notion of speed of displacement of the corpuscle of informational energy that is the photon. If time is immediate between two components linked by a substrate link, it will not be any more so between aggregates using sequences of links or the time will be the transfer between these sequences of links.

The generation of the Universe is thus a considerable process of emergences using a substrate of informational energy. It is necessary to explain how the emergence which generates and organizes its physical elements is realized, and for that we will have to proceed to modeling by specific multi-agent systems.

An activity information component is an element with autonomy, which communicates to form structured aggregates. It is active in the sense that it is made up of one or more specific informational fields and has an informational envelope grouping together its local fields, allowing it to define the characteristics of its current activity. It will have two different types of actions: either duplicate itself to create a new component, or communicate to commit to creating an aggregate with other components.

The structure and action of such an informational component comprising internal fields and an informational envelope is then as follows.

The structure and information envelope of an activity component

The informational component of activity is the basic informational action element of the informational model whose activation will engage the informational field that envelops it in action. It contains fermions according to its type, which are the basic elements of quantum particles. It is charged with informational energy by the substrate, either directly during its creation and actions, or by photon acquisitions. It is activated continuously, with regularity. It will produce and process generating information. It will have a structure with its own **informational envelope** that will envelop all its informational fields, unifying them, and it will be active with this enveloping field in the empty space where it is located. It can become inert depending on the information received, but it can always be reactivated by the communication actions of other components. This informational component, which has a major informational role in the Universe, will diffuse and regenerate by the act of its existence its own local information field in its generations of new components. It will generate components of activity or structure which, by their actions, will produce the elements of the Universe and of space.

The informational components of activity are launched initially from the generating component by massive generation and they produce, by their activities, a diffusion and a reification of their informational fields of envelopment. They can thus develop at each duplication action a similar or slightly different component at the level of quantum characters. When they produce a component of structure, an element of vacuum is produced forming a cell of the space of the Universe with diffusion of informational links in the substrate. It is therefore necessary to conceive that all the generated structure components constitute the constituent elements of the space which unify, which is easy to realize because they are informational fields which can aggregate without problems. There is a simple rule specifying the amount of empty space created. When a component generates another one, it has received an incentive piece of information from the substrate and it possesses or has received the informational energy to realize this generation. In this case, the generated component will have informational energy if it is of activity. If it does not have energy, it will be inert and will be of structure, it will be a new cell of space vacuum that will unify with the others of its neighborhood. The organizational law engages to generate a very important vacuum where the components of activity will be located and thus the component ensures the generation of many components of structure and far fewer components of activity, so that the organizational law generates a balanced universe between vacuum and elements of activity. We have thus a rather weak density of components of activity in the creation of the immense empty space of the Universe.

The first activities of the informational components and the initial Universe ball

– Initially, there is a creation of a huge number of informational components by the generating component, which form an expanding domain. This domain can be considered as the **initial ball** of the Universe and it is in considerable expansion. It would be the effective generating ball of the Universe, which forms a whole where all the elements of activity can communicate thus creating the notion of hyperdense informational link producing the initial space.

– Components can be generated as structure, creating new space elements. As activity elements are also generated, there will be two areas in the production ball of the Universe. There will be a domain where basic components of activity will be in a vacuum but fairly close, and there will be the periphery of this domain, where components without external neighbors will always generate informational space because they cannot communicate enough to satisfy the aggregation tendency. This amounts to an expansion of the initial ball with the domain of its periphery which has space to build up, because their elements have no close neighbors and are in the phase of space creation. This explains the continuous expansion of the Universe. If the ball was of radius R1, the new ball will be of radius R2 after a very reduced time and the domain of space created will be:

$$4\,\Pi(R_2^2 - R_1^2)$$

– Components of activity internal to the ball, having neighboring components of activity, will produce domains where they will be able to communicate and thus to associate, to aggregate, to form the new elements by compositions. The ball will thus become a set of domains having multiple internal forms due to the position of the sets of components of activity in its space and thus will be the space of the Universe with immense empty domains containing large domains of physical elements.

Thus, through the action of these informational components, we have a continuous increase of the initially created ball with internal domains containing a high density of activity components. The activity components located in dense domains will be able to switch to their communicative action state and form the quantum elements. The informational component will thus change its behavior.

The creative activity of the informational component of activity

The informational component of activity that can communicate with many neighboring components by the energy of the available informational space will modify its behavior, because it has in its informational field, in its structure, characters of basic quantum elements that will be the fermions, in the form of specific informational fields, and photons (see Figure 2.1). This spatial and informational envelope, which is its field of existence, defines its informational state by its form and its extent, and an informational contact with another field, at any place of the envelope, modifies it immediately, by instantaneous deformation, which is the effect of an informational communication. The component will communicate and associate with other components by informing of its quantum character to find a complementary component and it will put itself in state to create by association the atomic elements.

We will see in the agent modeling that represents the informational components of activity that it is possible to establish a permanent link between the envelopes of the two agents and thus between the envelopes of the two components, which will make these two elements entwined.

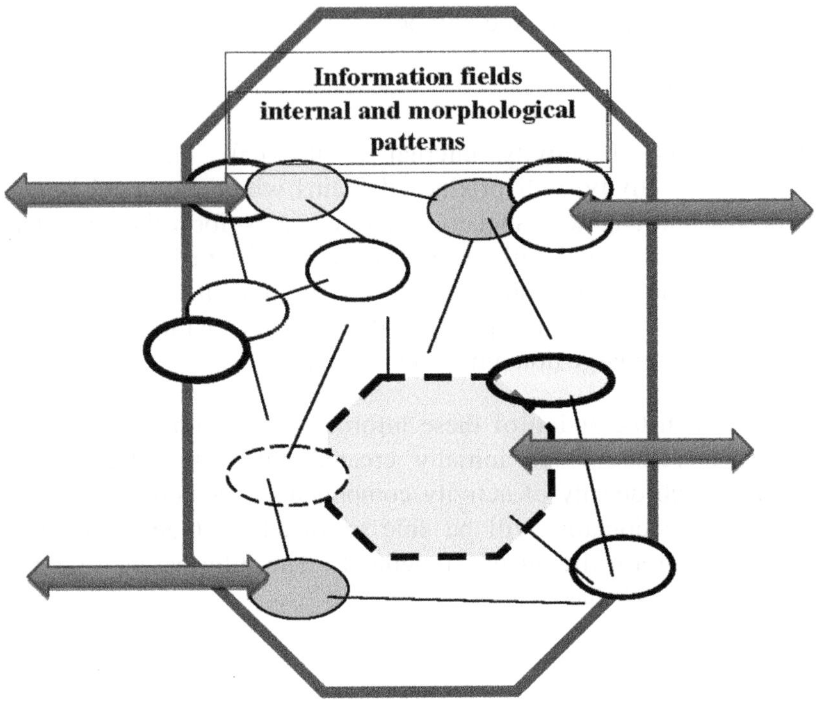

External information envelope

Figure 2.1. *An element made up of informational components forming its internal fields and its external envelope that communicates*

We can now specify what the basic quantum elements are in this model using informational components.

The quantum-type information components

In the model, each basic quantum element of an atom, that is, bosons, hadrons, leptons, nucleons and also photons, is specified by the specific characters of an informational component and these characters are located in the informational shells. A character can also be that of an electron, and this component will have to associate with a component possessing the basic quantum characters specified previously. We see there that any informational component has a role of generator of the quantum elements according to its state which is given by the general organization of all the components of activity. When an area of space is available and contains many neighboring activity components, all the components will pass into the role of activation of the quantum characters that they possess structurally. The activity components have been designed with these specific characters, with a good number of different and valid quantum elements created in the very many generations.

At the activation of these components to create the quantum elements, there will therefore be huge and very dense clouds of informational components of activity representing the basic elements of the atoms, which will allow innumerable aggregative communications via the informational space between all the well-structured components to engage their organizations. The informational envelopes of the components will attract each other very strongly and there will therefore be a very great commitment of this multitude of elements to confront each other to aggregate. It is quite obvious that the quantum-type components will generate all the quantum elements, which the builders of the Universe system have designed with the organization of the generating component and the organizational law as well as with the considerable generation of activity components of different and complementary types according to the impulses of the substrate.

We can then define the organizational- and informational-type of the Universe. The Universe system, at the level of the material elements, will be formed essentially of elements having informational fields which will organize themselves in more and more complex structures to represent the particles and all the physical elements. This is done in a space-time generated by the informational components in the phase of creation of the informational space, which will continue to be done at the boundaries of the Universe. There is, by the sending of information from the substrate, either an incitement to the creation of a structure component to make space, or the creation of an activity component to allow the creation of aggregates. This is defined in probability by the expression of the organizational law. The Universe has thus an organization in continuous expansion.

We can now define the spatial and temporal characters that exist in the Universe:

– Any active informational component is created as a local informational field by its envelope, which makes it exist. It will activate itself and continuously produce its current informational field thanks to its informational energy provided by the substrate. This is the basis of its spatiality. Any component is active by periodic instants, which founds its own temporality. It will activate its generation of structure components to create a new element of space of the universe, which will be a form available for the communications and usable as a spatial element. Thus, the space of the Universe is created, which is fundamentally informational and has an energetic dynamic to transmit fundamental information through the informational links between all the components of activity.

– The activity components will aggregate and form atomic and then molecular aggregates, each of which will have a spatial extent defined by its own informational envelope, using the available space.

– Finally, the aggregates will be structured in aggregates of bigger and eventually considerable sizes such as planets, stars and galaxies, and we will have there the notion of existence of material elements being able to move in the available space with informational envelopes of considerable sizes.

We can also define the notions of time, of duration, which are defined from the regularity of the moments of actions of the informational components.

We can specify three notions of time and duration in the Universe, which will depend on the situation of the informational envelopes of the considered elements and thus on the structuring of these elements.

The characters of time in the Universe

– The creation of the informational space by the components creating components in inert state is done at the speed of the activity of the components, and this notion of speed of their regular activity is given initially by the generating component of the Universe which fixes the frequencies of productions. The speed of action of the components to communicate or produce other components is thus fixed on the speed of the expansion of the Universe and it is the first notion of time that we can consider, the first notion of duration, which will be the minimal duration of an informational action in the Universe. It is, for us, immediacy.

– There is a second notion of time which is that of the transmission of the informational fields of regulation between the elements which constitute the aggregates of components, like the atoms or the molecules, by traversing the continuations of informational links, and also the transmission of information by information fields between aggregates, as is the case of the speed of the photon which is constant and which is the speed limit imposed on the effective displacement of an element of quantum-type. There, there is spatial displacement.

– There is finally a third notion of time which is given by the speed of movement of the material elements forming the very large aggregates as in the solar systems and the planets, and which depends on the size of the global informational envelope of these aggregates where the notion of gravity is exerted, based on the force exerted by the informational envelope of the aggregate to attract the elements which are in its close context and so that those which are there remain there by thus preserving the general structure. The time will then depend totally on the space to be covered and the energy given to the elements, it is physical time.

We can specify temporality in the structured sets of aggregates constituting physical elements. The notion of time must be based on the continuous and unifying organization of the informational envelopes of the great sets of aggregates which are being structuring. The time in these great aggregates is the speed of existence of the structures, and it is thus characterized by the **degree of informational organization** of the domain defined by these informational envelopes and where the aggregates constituting their elements in continuous structural communications are placed. For the spatial situation of any aggregate within an envelope, this gives its speed of organizational existence, by its internal and external informational exchanges with the elements of its context. This thus fixes the notion of a gap between the simple spatial immobility of a physical element which only organizes itself internally to exist and its movement of displacement in the environment delimited by the general informational envelope of the domain, which defines the maximum limit. Each organized element has an existence in a duration which is continuous allowing it to realize its structure and its deployment in this existential domain and which is of the same degree as its neighboring elements. There are then as many values of temporality and duration as there are significant organizational envelopes in the great aggregates of the Universe.

This important notion of the degree of informational organization of the elements of very important volume of the Universe will make it possible to define precisely the **notion of gravity**, which will depend on the extent and the structural quality of the organizational action of a large element girdled by its informational envelope and the vacuum.

There are thus many characters of space and time in the Universe, which depend strictly on the unifying informational envelopes of the considered elements. We will be able to specify that all the informational components behave in fact as agents of a particular type in a very particular autonomous multi-agent system which is self-controlled and which constitutes a description of the Universe.

2.4. The formation process of the Universe from the informational components

The Universe is a system in continuous self-organization which has components generating information space and quantum elements. We can describe in a general way the process of creation of the Universe system based on the use of organizational information:

1) There is initially a **generative informational component** created in activity, which is endowed with a considerable informational energy and which will generate a very large quantity of informational components of structure forming the spatial cells and of activity having each one a specific quantum character and whose composition will form the various atoms. This element will be reduced when the number of created components will launch the constructive generation of the Universe with a substrate having a sufficient quantity of informational energy.

2) A very large number of informational components are generated and made available, of which those of structure will produce informational space. The substrate must give informational energy to the generated action components so that they can be active. This is an organizational activity rule of the substrate, to generate a universe with a very important empty and available space. The activity components are endowed with the ability to act with regularity. These activity components will multiply by producing

others, according to their context. This initial generation launches all the basic elements according to a process of intense action with the creation of a deployment of the informational space at a frequency allowing the notion of temporality to be found.

3) When informational space is generated by containing swarms of activity components, the informational substrate will send them incentives to go into communication action to associate and create aggregates forming the first atomic elements. These constructive communications between the components will form atoms through finely structured aggregations.

4) Then, the communications between the structured elements of atomic-type will form aggregates of atoms which will be molecules, always under organizational incentive of the informational substrate.

5) Then, the aggregates of atoms and molecules will communicate through their informational envelopes to form more complex elements and organize themselves into denser and more structured material elements. The organization continues on a larger scale.

6) The material elements will form in spatial masses by large aggregations in the empty space; the masses will aggregate and form huge clouds of material elements.

7) The clouds of material elements will organize themselves into stars in galaxies, generating planets, and then organized life on planets.

It can be posited that the areas produced by the activity of informational components are not similar. There are components that generate a lot of structure components forming empty and available space, and there are components that generate a lot of activity components. This is because the generation of activity components requires informational energy, which is given by the substrate that may or may not give it to the generating activity components. It is a strict application of the organizational law which will produce material domains of various dimensions in the available spatial vacuum. Thus, we have the generation of a global whole where there will be swarms of activity elements in space formed by totally neutral structural components constituting a set of unified spatial cells. This is the key to the formation of a universe where there will be certain physical domains with very dense elements continuing to unfold their complexity, with separation of these domains by empty space made up of neutral sets of inactive structure components. The organizational law organizes the structure of the

Universe to apply in some places considerable increases in complexity, as will be the case with the creation of life on planets orbiting regularly around fairly stable stars.

This general process starts from a generating element producing a lot of informational components. These produced elements are strictly informational and autonomous and will not cease to generate new ones, of activity or of structure, without the incentives of the informational substrate. Such a generation process producing incentives leads to giving behavioral rules to these generated elements of activity so that they organize themselves in a coherent way by forming structures having stability. This has been done by the designers of the universal system who have defined a general organizational law of the system, which is in the substrate and which has informational elements of morphological control which it can exercise on the emerging components, in their communications and their available informational energy.

We can then specify the fundamental organizational scheme of all the elements of the Universe.

An organizational scheme of the formation of the Universe and gravity

Initially, the main activity of existence of informational components is to update their informational field based on the initial notion of action duration and thus producing structure elements becoming dynamic informational space and reconstituting their envelope to have informational energy taken from the substrate. They also produce activity components.

Then, informational components of activity will be able to aggregate, to form by grouping elements of atomic-type and then of molecular-type, and their organizational fields will change in nature in the new constructed informational aggregates. They will communicate by using their informational energy transmitted by the substrate, and there will be overall continuous use of this informational energy to form aggregates. The fields enveloping the aggregates will be attractive to the elements of their contexts, to produce larger aggregates and above all they will allow the aggregative structure that has been formed to be maintained. This aggregation force defined by the size of the informational fields of the aggregates will thus generate **gravity**, which makes the masses attract what is in their close context. The organizational rule leads to only two different actions: on the one hand, to generate informational space by creating structural components and then elementary components of activity and, on the other hand, to generate the memorial permanence of the association of the components of activity in the aggregates, by attracting other new elements to complexify the organization of these aggregates.

We can posit how the Universe was formed by its design system. We can posit that initially a set of processes of the informational components-type that constitute the basic elements of the Universe system are created and activated by the generating component using an available informational energy that comes from the creators' space. These informational components are structured by the incentives given by the organizational law coming from the incentives of the controlling elements of the substrate and they are all activated to generate a continuous expansion. In this framework, the Universe is the incessant informational unfolding produced by a system of generative informational energy available in its substrate.

In computer science, a way has been found to generate multiple processes active during the parallel executions of programs and running in the processors of multiple computers, with the generation of their communications realized by very powerful internal and external computer networks. The results of the programs are visible on screens in two dimensions. Here, in the case of the informational generation of the Universe, the processes are the activities of a considerable set of informational components that are informational fields, all of which communicate with each other and whose energy comes from an informational substrate. This activity would amount to unifying a considerable set of processes representing the components in activity with the use of an equally considerable memory formed by the informational envelopes of the components and the informational links of the substrate allowing the communication between components in the space of informational nature. The image of these forms of informational activities in the computer-based reality would amount to unifying the generators of electric energy, the processors, the memories and the networks in three-dimensional space of action. The Universe could thus be considered as being a considerable set of processes of continuous calculations and communications with an expansive dynamic memory, the whole generating an autonomous system organizing itself by its actions of communications. How can we create such a system?

There, we can approach the problem of multiple universes or multiverses (Luminet and Lachièze-Rey 2016). However, we must approach the problem of multiple universes from another point of view than the idea of an infinity of parallel universes generated during a creation. We can posit that the creators of our universe have conceived it from a generating system producing innumerable elements of the informational components-type provided with the right rules of action and communication, to generate a formidable space-time expansion which would be a form of the dynamic informational actions memorized and preserved by the system in action. This system of creation of our universe would be generated in another universe and these two universes would be parallel and not directly communicating. We can thus pose that for the designers, it was possible for them to generate a system of creation of another universe starting from the design and the activation of a system generating quantities of informational components, the whole in connection with a substrate supplying continuously informational energy which would be the force of action. It was possible for them to activate the entire hyper-massive informational system by an expansion action that was carried out outside their universe, because the informational energy of the two universes is different. How to proceed to this action of creation by producing it outside the generating universe?

We must assume that any universe is a system of elements using informational energy, being an organized expansion from its initial state with the creation of a set of informational components in action to create a proper space-time. It is necessary to pose that this own space-time of totally informational nature is placed outside the space-time of the system of conception of this universe.

We can think that it would be possible, in science, to conceive elements of the informational components variety possessing the good characters of communication, able to generate dynamic memory using informational energy and to make them activate themselves in the form of autonomous informational processes. We will show in the following how to program a universe. It would be necessary to conceive these elements from the quantum level by going to the substrate level, by specifying their characters and making them multiply very strongly. It is also necessary to define and use the informational envelopes of the atoms and molecules of the creation of this system. Then, by using layers of photon organizations possessing energy in quantum computers, it would perhaps be possible to launch the whole of these fermion-type informational processes. This dynamic whole would be autonomous, of very small size and would create its own space-time independent of that of its creators. We can perhaps posit that this was realized a first time on a very small scale, in the creation of a very small universe. In this initial universe, at a certain very energetic point there was a global explosion and then the generation of another universe with the available energy which was of bigger size and so on, to end up with the formation of greater and greater universes, like ours, which is of considerable size.

There is thus a system of design realized in a universe and there is the launching of this system of design in the form of a self-controlled generating system functioning in an autonomous way at its own informational frequency and generating a universe. We describe the algorithm:

Algorithm for generating a universe

1) The design

1.1) To have in the design system a considerable flow of informational energy that will be in the form of available informational fields.

1.2) Build a system that generates three types of information processes:

– The structure component-type process that can form a local field of spatial vacuum and that can morphologically observe the state of the active activity components.

– The activity component-type process with fermion definition characters, communication initiation characters, communication reading and interpretation characters.

– The process of informational link-type memorizing a generation or a transmission of message from component to component and capable of transmitting informational energy, which will form the informational substrate of the system and which will allow for realizing the morphological observation of the activity components. It is the realization of the data for the control.

1.3) To have a global process coming from the design system to distribute informational energy in all the links of the generated system substrate.

2) Self-control incentive

Process of observing through the substrate links to represent the morphological state of the activity components, morphologically analyzing these sets of activity components and sending incentive information to put the organizational law into action.

3) Launching the Universe generation system

Condense all the processes formed in the generating system by making them active at different informational frequencies than the design system and continue to launch informational energy onto the forming substrate.

Therefore, we consider that the Universe is the most complex autonomous and self-organized process possible, generating itself from a singular informational element, functioning for a long time and destroying itself after a certain time, and then reconstructing itself. We consider that the space of the Universe is absolutely not something totally neutral and available, which comes from nothing. It is a space on a substratum of informational energy, created with active basic components which will allow, by multiple constructive incentives given to these elements, that their organizations are realized and that the whole becomes a system in continuous organization at all its scales. This posits well that the generation of the Universe was conceived and that it continues its realization under an organizational law.

We will see that this informational model forming the physical elements can be represented with informational agents, by analogy with the organizations of software agents used in computer science.

3

An Agent Model to Represent Informational Components

We will see that we can understand the informational components as informational agents, of a type quite similar to the software agents defined in computer science, which will allow us to refine the modeling.

3.1. Informational and control agents representing the components

Work in Artificial Intelligence has very clearly defined what a software agent is. The definition that characterizes the autonomy and the self-modifying role of an agent is the following (Ferber 1999):

– Any software agent is an active computer entity, which is a coherent set of programs that can run and calculate values, by the choice of programs operated according to parameters.

– Agents are entities considered "social" in the sense that they communicate with each other using a simple, fully determined, precise and unambiguous communication language.

– They are reactive entities, which have a perception of the state of their environment and the modifications of this state leading them to adopt a certain behavior, that is, to engage certain actions by their programs. They are built to engage these actions within a time frame consistent with the dynamics of the environment.

– These entities are mainly **proactive**, that is, they can initiate actions without necessarily being led to them by explicit external solicitations. The agents thus have a behavior with an autonomy of decision and action. They can thus refuse another agent's request for action, even if it is imperative.

– They are evolutionary entities: their behavior leads them to modify their goals and the means to reach these goals, to rewrite their programs by themselves and to modify their program choice parameters.

– They are entities forming a homogeneous algebra: an agent can generate several others, of appreciably different aspects, and an aggregation of agents can merge to become a new agent whose structure and actions are a synthesis of the characters of the designer agents.

We propose considering the informational components as strictly informational agents, all of them organized in innumerable local multi-agent systems constituting the Universe system in organization. We will thus define a structure of the agent-type for the informational components, by always keeping the notions of informational fields and informational envelopes and by not using the notion of program, which does not exist in the basic elements of the Universe which is an emergence on an informational substrate.

The structural characters of an **informational agent** are then the following, which will be morphological characters of the unifying informational envelope of the component it will represent.

The informational agent representing an activity component

– The category of the informational agent will specify, according to the morphology of the informational envelope of the component that it represents, the type of quantum element that it will be able to represent. We can thus consider that the agent has an informational envelope which constitutes its structural memory.

– The state of the agent is to be active or to be in a reduced state, on standby and therefore inert.

– The activity rules of the informational agent are:

1) General rule of duplication of the active agent with the creation of another agent that is either active or reduced: creating a reduced agent simply constitutes an empty cell of information space allowing the

application of the general organizational rule of aggregation and this agent then represents a structural component.

2) General rule of action of communication of the active agent and search for other agents to join: choice of information in the form of specific information field sent to receiving agents and if there is a good answer to its message, there is treatment of the received information, corresponding to the deformation of the informational field of the component envelope. This communication is thus of the transmitter–receiver type. It is done at the maximum speed for the notion of speed used in the Universe system and we consider it as immediate. It has as a natural continuation rule 3.

3) Rule to join with another agent to form a new element by unifying its rules with those of the other agent and adapting them to the new structure of the newly created agent.

4) Specific rule of possible change of the strictly autonomous status of the agent if the communication with another agent is shared in a constructive way: physical association of the two agents, with the unifying creation of a new agent containing both of them, with structuring, with new action rules corresponding to the new structural state.

5) Specific rule for switching to the inactive state, if an imperative informational message from the environment agents requests it.

– Meta rule for choosing one of the rules of action according to the current situation to adapt to the organizational activity of the environment.

– Meta rule to commit the union with another agent to being systematically constructive.

– Meta rule to receive information from the global context of the envelope of aggregates of agents, and action impulse to organize the context, in order to make it apply the organizational law by modifications in the application of its rules.

– Regular energy intake in the informational substrate allowing all rules to be available to activate and the agent to have an active state.

– Elementary history of the agent's main past actions, memorized by the morphology of its informational envelope.

We can then consider an informational activity component as being structured with action rules like the rules of an informational agent. All the rules are informational characters, that is, characters of morphological type given by the local internal fields of the component and by its informational envelope. Such an informational agent is thus a strictly communicative entity that acts with its own rules, that can generate other similar agents according to its state and that of the context, and that communicates a lot. A structural component could also be represented by an informational agent, but with very reduced activity rules representing simply the informational action of the links between the local substrate and the component. We will thus use another type of agent to represent the action of the substrate: the control agent.

The two types of agents corresponding to the two types of components

The **informational agents** that we will use will represent the activity and structure components, and the links of the substrate will be represented by **control agents**. The agent model is thus representative of the activity of construction of the material elements of the Universe, and not of the generation of its space nor its temporality.

There will be basic informational agents representing the components of activity with the quantum-type atoms, and then there will be creation of more complex agents, formed of aggregates of these basic agents and which will be agents of atoms and then molecules, because a union of agents is always an agent. Thus, we have an effect of the organizational law.

Consideration of the organizational law with the control agents

All the entities generated in the Universe system are wrapped by informational layers which are local to them and which allow communications. In the agent approach, we start from the basic agents representing the informational components, which aggregate to form all the other agents that will represent the physical elements. Thus, all the elements generated at all scales, which can be considered as very strongly communicating agents, exercise the rules they possess and which define their behaviors according to their environment and the messages coming from the substrate and triggering their types of activities. This constitutes the continuous application of organizational rules at all levels by a perfect multi-scale communication and it is indeed an application of the organizational law. We can therefore consider that in the multi-agent system, there will be an internal organizational control system with **control agents** adapting to the organization of informational agents and operating by morphological and semantic analysis (Cardon 2005). These control agents, which will explicitly represent the action of the substrate elements and the structural components, will send solicitations to the groups of informational agents according to their conformations and states, which represents the organizational law. They are the correspondents of the collective activities of the components' informational envelopes.

An active informational agent will thus have two main phases of activity where it will successively apply its two general behavioral rules, the first being reproduction and the second being unification with another agent. The first rule is simply reproduction, and in its state of construction of elements corresponding to its second rule, its action will be to form aggregations with other agents following communications of an organizational nature. Its behavioral rules are reified at the component level in the morphology of its information field located in its deployment envelope which is its informational extent. This informational agent will realize the effective memorization of its actions by morphologically modifying its informational envelope. In computer science, the agents use local memory which is taken from the general memory of the system which activates them. In our case, the informational agents generate the memorization of their actions carried out in the spatial extent of their informational envelopes. These memorizations are essentially carried out in morphological form, modifying the frequencies and the energy of parts of the spatial extent of the envelopes. Any informational agent has thus a spatial domain which will systematically organize itself with those of the other agents to constitute, by all the communications, the whole which will form the material elements of the Universe system.

This agent has a specific structure and a membrane that is valid for its system of existence and communication, allowing it to receive and transmit information to other informational agents. Any information that is sent to it is an informational field that has a morphological form indicating what the communication means, and there are the following three main forms:

– Communication of linear-type with one or more neighboring agents for a union to be carried out in order to generate an aggregate which will be a particulate element respecting the rules of its creation. This is how atoms and molecules will be made. If this union is carried out, there is then unification of the informational envelopes and thus creation of a new informational agent in the union of the spatial fields, having new properties which will define its rules of behavior.

– Indication to the agent receiving an information to stop being in a constructive stage, to eventually return to the state of an informational component of structure that will generate informational space again. This incitement is most of the time sent by the substrate which will be considered as composed of **control agents** linked to all the informational agents and which will be able to send these messages.

– Repulsion of the context by the agent, refusal to interpret the information received and actual displacement to change context.

The notion of control agents is important in the multi-agent model because it specifies that the informational substrate is both a set giving basic informational energy and giving incentive control-type information to all agents to apply the organizational law. These control agents will represent the organizational actions of the substrate. They will be of the same type as the informational agents with simple specific action rules that will be as follows.

Rules of action for a control agent

– Locate and analyze groups of active informational agents and analyze their types of behaviors to see their importance and the consistency of their states and activities. This is a morphological analysis that is also partially cognitive.

– Send a constructive organization message to the informational agents of a well-identified group, which is an incentive to form an aggregate of agents that will be represented by a new agent.

– Send a message to an agent to place it in an inactive state or to resume activity if it was in an inert state.

The communication of informational agents by sending messages in the information fields of the connected agents is done practically immediately by using the network of informational links of the substrate, because it is a question of almost immediate morphological deformations of the enveloping fields, of modifications which are done at once on all the active field. The communication can also be realized by an active informational link which becomes permanent between the informational envelopes of two agents, which are thus united in their activations. It is this property that will define the notion of **entanglement,** which has been verified in quantum mechanics (Aspect 1976).

The entanglement of two informational agents

We consider two informational agents that are in a very strong union through informational fields that are in the form of an active linear link between them, a link that contains informational energy and that thus conveys information in continuity. In fact, a control agent links them permanently. Their informational envelopes are then in unification for these two agents which are spatially separated by a physical distance but united at the informational level. The two informational envelopes are separated in space-time, but establish a continuous informational link between them, which will allow their permanent communication. If one of the two agents is altered and sees its informational structure modified, then there is immediate transmission to the other agent, without relative duration of transfer for the notion of space-time duration. The characters of this information carry out the morphological modification of the informational envelope of the altered agent and the other agent carries out the same modification of its envelope, at the same time. This is how we have thought, in physics, of a particle positioned in two different spatial locations, as, for example, an electron that would be in two locations in space, but we are talking about two particles linked by a permanent informational link between them, establishing their behavioral identity.

Let us note that the formation of the aggregates of informational agents that will form all the atoms will engage each time to generate a new informational envelope for the new agent that will be the union of the local envelopes of its component agents. This envelope will lead to new behavioral rules that will specify what the new aggregate agent must do, defining these new rules as a synthetic union from the rules of the generating agents. This is indeed the effect of the organizational tendency of formation of the Universe: as soon as we move on to elements of higher scale, the constituent elements form a new unifying informational envelope that defines the new rules of behavior.

We can then define how atoms and molecules are formed in this agentic system of informational components.

3.2. The generation of atoms and molecules in the informational agent model

We are in the stage of description of the model where we consider that space is available and that it is now a question of generating physical elements of all sizes. It is thus a question of specifying with the informational agents how the atoms are formed, then the molecules and all the chemical constituents generating the material elements of the Universe. The informational agents will permit the accurate representation of these formations, the constitution of the physical elements of the Universe being apprehended as a self-organizing process managing an immense informational multi-agent system. The element which will allow the formation of the atoms will be the informational agent which represents an informational component of activity. There will be a union of informational agents representing specific elements to form a new atom agent. The basic rule used is that a union of agents can easily form a new agent.

Let us recall the characteristics of informational agents.

The properties of informational agents

The basic elements of the constitution of the Universe system are the informational components of activity, and these elements are constantly activated to update their informational envelopes and then to form structured aggregates. They receive their necessary informational energy from the informational substrate. The structural components cause the spatial deployment of the Universe. The informational agents, by contrast, can move in the space considered as created by emitting and by receiving the communications generated between agents. They then move to the stage of aggregate generation and form atom-type elements which will have their own informational envelopes. The informational agents are in relative mobility, and this mobility gives them the means to act and to communicate. They will thus be able to pass to the level of the effective creation of atomic elements by sending information in their contexts to cause aggregations.

The basic informational agents represent activity components that have internal fields which define the elementary quantum particles that are bosons, hadrons, leptons and nucleons. These will be specific internal characters of the informational agents. Each quantum component forming an atom, the nucleons (protons or neutrons) and the electrons, will be thus an autonomous and active basic informational agent.

We therefore consider in agent modeling that space is available for their actions and that informational and control agents will generate agents corresponding to all aggregates, from atoms to molecules, to all physical material elements and then to all living organisms.

We have then the following property: the basic quantum elements are informational components that we represent as informational agents; these agents have the property of representing in their envelope the characters of the basic quantum elements that they possess.

We will give two examples of the formation of atoms, the hydrogen atom and the helium atom.

3.3. The formation of a hydrogen atom agent with informational agents

We will show how a hydrogen atom is formed. Hydrogen atoms are considered to be the most numerous in the Universe, with a quantity that is evaluated at 92%. The very high density of informational components in the initial domain of the Universe will thus lead to the production of mainly hydrogen atoms which are of simple design, because the contexts containing multiple informational components of activity follow the organizational law and will engage a lot of aggregations which will be structurally the simplest. Let us now use informational agents.

The production of hydrogen atomic agents will lead to the production of a spatial domain where there will always be informational agents in the initial phase, then hydrogen atomic agents, and where there will also be groups of hydrogen atomic agents which will then evolve, leading to the formation of more complex ones and producing aggregates which will be the future molecules. This is the application of the organizational law, where the behavioral rules of the agents are applied, making them generate, complexify and organize themselves.

There are swarms of informational agents, many of which have the category of being a proton or an electron. Let us consider two informational agents, one of which has the category of being a proton and the other of being an electron. These two agents will associate due to their systematic communications forced by control agents, which will be the intense communicational state of the beginning of the formation of hydrogen atoms.

We will specify the structure of an informational agent by describing a proton agent. Its structure allows the following actions, which have very simple algorithmic characters:

– The considered informational agent has the category proton, with the precision of the characters of the proton in its informational wrapping memory.

– Agent status: active.

– Communication action with the other agents in its context and search for an electron agent. No link with another agent if it has not found an electron agent in its context. An electron agent may equally have initiated a communication as a requester and the proton agent may have received it.

– If the agent encounters an electron agent in its context, then there is strong binding communication between the two agents, then association action if accepted by the proton agent and the electron agent, which is the normal rule requested by control agents.

– Realization of the union: change of category of the two associated informational agents of types proton and electron, use of the characters of the proton agent and the electron agent during the association. Spatial union of the two agents and formation of a specific informational envelope for the new agent which will then be of atomic-type: major action of the two envelopes of the proton and electron agents unifying to define the new characters of the new atomic agent with a structure using those of the two designer agents. A new action rule for this new hydrogen atom agent will be to look for a neutron agent in its context to eventually form another atom agent. The new envelope has a spatial character, with a larger extent and the two designer agents have become internal elements in the form of active informational fields.

– Minimal memorized history of the past actions of the two agents and their transformation by the association, stored in the spatial envelope of the new atomic agent.

The structure of a proton-type informational agent thus has a principal rule of action consisting of seeking contact with an electon-type agent, and reciprocally, an electron-type agent must seek contact with a proton-type agent. It is a question of these informational agents seeking to associate, to generate, for example, a hydrogen atom agent which will have a character of stability and which is a very simple element to carry out in the organization which is generated. But each of these agents enters into repulsion if it finds an agent with a type with which it should not associate: the agents are created to unify with agents that we define as being of **complementary type**, such as proton–electron, which is a rule of their behavior that is also driven by the control agents. This rule clearly indicates that the system is not made at random, but follows very precise constructive rules that are in the informational components and therefore in the agents.

There is in the substrate, a set of elements that will be control agents, which analyze the morphology of the set of informational agents that are activated, forming space or active informational agents. They look for the characters of this morphology to send messages of constructive incentives in the groups of informational agents that allow it. This eliminates the incoherence in the action of deployment of the informational agents. They are the ones who strongly solicit the proton and electron agents to communicate and then to associate. To do this, they give informational energy at the time of the association and then they give it to the atomic agent created.

Action of substrate control agents on informational agents

– Identify highly active groups of informational agents.

– Identify the characters of informational agents in these groups and the general character of the group. They have knowledge about the categories of informational agents.

– Look for agents in the group who have complementary types and can be associated.

– Send incentive messages to these pairs of complementary agents so that they send themselves association messages and associate by forming a neutral element and send them photons giving them informational energy.

– Follow the behavior of agents that have become atomic.

– Send informational energy to the generated atomic agent so that it becomes active.

These actions of control are continuous and permanent and they are incentives that can be refused by the informational agents, according to their context. We begin to see that the central point of the Universe system is the control exercised by the control agents of the substrate to achieve all the organizations. It is an incentive control, and a major one. It is multiscale and allows the observation of the morphology and semantics of groups of agents and groups of groups up to groups of atomic agents. These actions clarify that the Universe system is generated on the basic elements, the informational agents, with permanent self-control and is not produced by simple chance in a hypothetical neutral vacuum. We have already seen that the substrate that operates the control and the sending of informational energy is the dark matter of the Universe, and in the agent model, the substrate is constituted by a multi-agent system controller.

The complementarity of informational agents

The notion of complementarity of the agents is very important and is in total coherence with the organizational law. This complementarity was produced by the way the informational components were generated by the base component and then the component generations. The control agents of the substrate give information to the informational agents that activate some of their rules when they consider that some agents are good for them in the formation of an aggregation and others are not to be considered. Some informational agents are considered by the substrate control agents as associable and others as non-associable, which is a basic clarification for the organizational law that the informational substrate control agents define. This comes from the generation of informational agents of different types and containing complementary elements in order to produce, subsequently, the stable atomic elements. Complementarity means that the association of complementary elements will have stability, which is expressed in physics by opposite charges for the elements considered, which together form a neutral state. The formation of stable atoms and molecules is therefore a planned organization, which can and will take place. The chance is only in the speed of generation of the informational agents and in the speed of realization of the aggregations, which will create groups which will not necessarily all be homogeneous.

The action process of an informational agent that follows the organizational pattern given by the substrate control agents is as follows. The proton- and electron-type agents seek each other out by being prompted to do so by their activated action rules and each sends a simple cooperation message to the agents in their contexts asking the receiver if it is the right complementary type for an aggregation. If a proton-type agent communicates with an electron-type agent, then they both immediately enter into strong cooperation and they aggregate. Immediately, they unite their specific fields and informational envelopes to generate a new one with a new form and they change states, which their functional systems can do by nature. A new atomic-type agent is formed which has a different informational envelope and whose new rule, which will be its main rule, will be to look for another agent to unify with it in order to constitute a more complex and always stable element. As soon as the informational agents enter in aggregation, they form a new atomic-type agent, with a new informational spatial envelope. This new agent formed by the association of the two elementary agents will have stability, continuously activating its specific internal fields that link these two types of former internal agents within its structure, making it stable.

We have here, in the realization of this aggregation of agents to form an atomic agent, an energetic property of great importance.

The informational energy and the physical mass of the elements

The formation of an aggregate of informational agents forming an atomic agent gathers its local components and realizes a new informational envelope that establishes its existence. In the formation of this informational envelope, there is the sum of the envelopes to achieve their unification in the new envelope. There is also the sum of the informational energies of the envelopes of the constitutive agents. This means that the informational energy expressed by the new envelope to act in communication, to memorize messages and to form the expression of the characters of the atom will be able to do so with more intensity. The atomic element will therefore have more informational energy available than the agents that made it up. It has gained in capacity of autonomy and action. There, in this fact of being structured, stable and having more informational energy, will reside the usual notions of physical energy and mass.

This atomic agent formed by two complementary informational agents has stability, and its internal structure is composed of fields that define its energy value by activating and communicating to ensure its coherence and stability. The notion of informational energy of the atomic agent will be the continuous interaction between its internal informational components, which realize the coherence and stability of the element and which unfold in the enveloping membrane. There, the notion of communication time will be able to change. The atomic agent will be able to communicate by sending photon-type fields moving at the speed of light, thus using its energy and that of the substrate, but the path of the photon will generate memory energy in the informational space. The atomic agent will be able to gain energy by absorbing fields, for example, of photonic type. There is thus a structural change with respect to the initial informational agents, we have entered the field of the effective generation of the structured elements of the Universe.

This hydrogen atomic agent may seek to aggregate with a neutron agent to form an isotope that still has stability, or it may associate with many other hydrogen atomic agents to form a large set of similar atomic elements, which is a tendency of the organizational law that drives them to perform these actions.

The general structure of an atom-type informational agent of the with its new rules deduced from those of its component agents will be as follows.

Structure of an atomic agent

– Category of the atomic agent: informational clarification of its atomic category, regrouping the elements that compose it.

– List of complementary types of agents with which to join.

– State: necessarily active if it is stable and dependent on its energy context.

– Continuous reception of information from the control agents, with consideration and possible modification of the current behavioral rule.

– Continuous action of the internal informational fields, whose informational envelopes have become internal fields between basic agents to maintain its structure and general envelope.

– Agent energy and context of the available information space with possible displacement to another context.

– Rules of action for communication with complementary agents accessible in the context.

– Rule of association: if meeting with an agent compatible with its evolution and in a state of agreement, then the action of association of these two agents, which can be an increase in the number of electrons or basic atomic elements, or an association with another atom-type agent to form a group of similar elements.

– Displacement rule: if elements incompatible with its structure are encountered in the context, then action for context switch.

– Location of a minimal history of past actions in the envelope morphology.

This atomic agent thus stabilizes its structure, keeping it in informational form, which is the central property of memorization of structures having stability and created according to the organizational law. All these rules will be expressed, in the agent, by the morphology of its informational envelope and by the internal element representing its type, which is an internal field and which will be the element deciding the choice and the action of its rules.

We can now define more precisely what the atomic agents will be spatially and temporally. They will know an informational envelope representing spatiality:

– The informational envelope of an atomic agent is the organized synthesis of the envelopes of the informational agents that constitute it. There are still types of reduced internal envelopes for the elements composing the atomic agent, but it is the new global envelope that identifies it. This new envelope defines its own spatial domain, which is composed of specific forms indicating the roles of this new element created by an organized association. It specifies its structural stability and its tendency to form aggregates or a new more complex element.

There will also be a **notion of time with the atomic agents created**. The communications between the informational agents which are elementary morphological fields in the Universe system, are immediate according to the notion of time of our universe, through meeting of informational fields produced by their envelopes. The atomic agents communicate by sending specific informational fields in the informational space initially created for the informational agents. The spatial envelope of the atomic agents, which defines their own domain of existence, can use the available space, that is, the available informational domain, to send informational agents, such as electrons or photons. The displacement of an atomic agent is an operation that takes place in the available space, and it is done at a speed depending on the energy of the atomic agent in the available spatial domain that allows the displacement, which is a transfer of informational envelopes. These displacements will be done at a certain speed; they will thus take a certain time which will always be lower than the speed of light. Time has become, for these atomic agents, the duration of their physical displacements and the displacements of their elements in the informational space.

We can define a **character of the organizational rule** which will be production by similarity. In the elementary space with countless informational agents, when a new atom-like agent has been generated by agents and is very stable, then it is relevant for the organizational rule to have this type of generation apply to many informational agents of the type that formed it and are available in the space. The control agents thus operate the elicitations with the rule of repetition of a successful aggregation.

In this stage of the creation of the Universe, there are thus associations generated by the informational agents according to their rules of action, with formation of atomic-type agents having new spatial envelopes giving them existence. This can be realized thanks to the considerable density of informational agents in the initial extent of the Universe and to the incentive action of the control agents of the substrate. The generation of the informational agents from the generating component defines, by the reproductions of the informational agents, a domain with a lot of informational agents having characters which normally allow them to realize the aggregations in atomic agents. Their rules of behavior are applied so that they generate atomic agents when the neighborhoods allow it by containing complementary agents.

3.4. Formation of a helium-type atomic agent

In the organizational domain of the early universe where there are many simple atoms like the hydrogen atom, new types of atoms will form, as well as aggregates of atoms leading to the formation of molecules. The Universe system is initiated in its construction by the continuous prompting of the substrate control agents.

Helium atoms are much less common in the Universe than hydrogen atoms, representing only 7% of the atoms. But they are simple and can be directly formed with the informational agents.

Formation process of a helium-type atomic agent

1) The context of active informational agents is dense, and many agents with neutron, proton or electron types can communicate in the same context.

2) The substrate control agents simultaneously induce neighboring proton, electron and neutron agents in a domain to enter into communication in order to aggregate to generate a new stable atom. Two proton agents in the domain under consideration then enter into strong communication in their context to associate simultaneously and two electron agents that are in the same domain enter into communication to associate with the active pair of proton agents.

3) These two pairs of agents that are in association then enter into association all together and receive photons from the substrate control agents so that they have the ability to perform the association. Their communications change the morphology of the informational envelopes of the four agents. But they are not yet stable enough and their organizational tendency will make them immediately solicit two neutron agents to produce a stable atom, which will also be an incentive of the control agents.

4) In this context, neutron agents are then very strongly encouraged by the organizational tendency of the control agents to enter into strong associative communication with the proton and electron pairs and eventually they associate all six.

5) There is then a strong association of the three pairs of agents: proton, electron, neutron, to form a new agent of a helium-type atom which will have an informational envelope unifying the six envelopes of the elements and establishing its stability.

6) The internal fields of the new agent are formed to ensure the existence and stability of the structure of the helium agent.

7) The envelope for this new agent specifies its characters and communications.

The formation of this helium-atom-type agent is a very important step, because it associates in total six basic informational agents which were all in good spatial proximity. It is a doubling of the formation of the hydrogen atom, with in addition the use of two neutrons. So, we see that the organizational law used a property to complexify the generated elements.

The organizational property of creating large atoms

Substrate control agents can use a simple rule to increase the complexity of elements, which follows the organizational law. It is a question of considering what has already been realized – here, the hydrogen atoms of simple structure – and to generate a new element that will double this structure, realizing a new stable element with double the components of those of the used elements and the elements stabilizing the new structure. To consider well that this construction follows the organizational law and does not come from the chance of the communications of the components, it will be necessary to add two new elements of neutron-type in the construction to have a strong stability and thus not to leave to the elements of simple structure, like the hydrogen agents, the autonomous possibility of doubling by putting themselves together to form unstable pairs. The control is thus major and always exerts its constructive law.

It will not be easy for the organizational tendency to go towards the formation of more complex atomic agents, possessing more components and remaining stable, because this tendency of the organizational law is very much in accordance with the general state of the considered field; the elements are all in coherence. It will be necessary to wait until there are very important clouds of hydrogen atoms with helium atoms to pass to the generation of aggregates of more important size and to the formation of other atoms.

The information fields of the envelopes of the informational agents have played a precise and very important constructive role in the formation of these new atoms. They have allowed the construction, with strong communications and using the elements of proton, neutron and electron types, of atomic elements with stable structures at the level of their charges. The organizational tendency will continuously generate new stable elements by balancing the charges of their internal fields.

In the examples given, a hydrogen-atom-type informational agent has just been created from component-type informational agents, and this will be very common in the considerably sized set of initial informational agents, which have as elemental types neutrons, protons and electrons. There are also photon-type agents that will be used to directly exchange informational energy between the elements, which is called transmitting charge, which is mostly structurally modifying information. There will be thus, by the generation process of the Universe, creation of immense clouds of informational agents of simple atom types, which will then be able to internally combine to constitute the molecular physical elements of the Universe, and then to generate new agent atoms in the deployment process which is always in potential transformation.

One can thus state the reason which will bring into existence atoms and then molecules.

The rule of constitution of the organized elements of the Universe

The informational agents have an informational envelope that systematically leads them to communicate. Their production actions lead them to massively duplicate themselves and to form dense clusters of informational agents, which will engage the substrate's control agents to be able to constitute atomic agents. The new atomic agents will have an informational envelope that will characterize them and that will specify their new behavioral rules, and then they will form aggregates of atomic agents to constitute molecular agents. This is the action of the organizational law that encourages all the elements to organize themselves in a continuous way so that they produce aggregates and then aggregates of aggregates.

Some other atoms like beryllium and lithium can be formed in the clouds of informational agents and hydrogen agents. For the other atoms, it will be necessary to form the considerable aggregates that are the stars, where new more complex atoms will be formed in the thermonuclear reactions of the stars. The organizational law is therefore the formation of clouds with the selection of elements forming aggregates defining new elements.

The organizational law has done its work of organization in the Universe formed by innumerable complementary informational components, close to each other, producing atomic aggregates. The creation process of material elements has thus begun.

The organizational rule defined in the substrate and applied by the control agents commits to form first dense clusters of informational agents, and then in these clusters to generate aggregates that will be the first hydrogen atom agents and also some helium atom agents. The deployment of all the basic elements is not done by simple chance but it is morphologically managed by the organizational law. Thus, the organizational law will continue its rule of formation of more and more complex clouds and aggregates by taking many hydrogen atom agents to form molecule agents composed of these hydrogen atom agents. The formation of clouds of elements will continue to form, in the space defined by all the agents, dense clouds of atoms and molecules which will be the first systems leading to the formation of starry clouds. The action of the organizational law thus carries out a succession of generations of more and more complex clouds and aggregates of elements, distinguishing for the aggregations the good choices of elements.

We can now proceed to the modeling of the generation of molecules by atomic agents.

4

The Generation of the First Molecules in the Agent Approach

We consider that we have, at this stage of development, a very large universe where the informational components forming dense clouds continue to generate innumerable clouds of hydrogen atoms, helium and where the active elements seek to aggregate. There are very large sets of almost empty space which only serve to locate the aggregates of elements. So, we have a system in deployment in a state that is going to transform by strongly organizing itself by producing massive aggregates, where many atoms are going to aggregate into molecules and where the new informational envelopes of these aggregates of molecules are going to form structured systems, where new atoms and new molecules are going to be able to be created. We are in a system experiencing very strong generation of complex elements.

4.1. The informational characteristics of the system forming the molecules

We are in a system formed by swarms of atomic agents in movement and in very important communications, where the notion of strictly local action will be replaced by the notion of organization on a larger scale. There is the effect of the global tendency to form groups, which is the great organizational hypothesis that we pose.

The global trend in the formation of molecules

In the system constituted by a set of atomic agents in action of informational communication, a strong tendency of the organizational law will engage the actions of organization of these atomic agents. The atomic agents located in dense contexts will all tend to form specific aggregates that are very well structured, which will be molecules. This tendency will be exercised by the communication between the informational envelopes of the atomic agents, which will systematically seek to aggregate to generate the new agents which will be molecular.

We can then pose a hypothesis on the very strongly informational character of the Universe.

The global information network representing the trend system of organization

At the stage of development of the Universe that we consider, we see it as formed of innumerable informational elements localized in a space that is the reification of the communicational memory containing its innumerable atoms, and we posit that the informational envelopes of these atomic agents will generate a **global informational network** that will be in the form of a union of dense graphs linking all the communicating informational layers, leading to the organization of all the agents in order to make them eventually change their structural scale. We have already presented this network with the network of informational links of the substrate. We thus consider that the multi-agent system representing the elements of the Universe has an envelopment formed by the network of informational links between all the informational agents; it is a hyper-complete graph which allows the control agents to exert everywhere its **incentive control of the actions to be carried out**. This network will be the structure that will allow for realizing the incentives of the control substrate. We will have therefore an informational network engaging the informational agents of atoms to proceed to local aggregations to form molecule agents, whose new specific informational envelopes will then form a new type of element of the informational network, and then in this local network, the molecules will associate and will form small molecular material components and will generate another informational sub-network.

We thus have a universe organized by components activating themselves in a set of multiple dense informational networks, allowing them to generate material elements and memorizing their structures to then develop them. These organizational networks operate with organizational tendencies which are information of the control substrate applying the organizational law by sending messages of composition. We will see later that the informational networks of the material components will organize themselves on Earth to form very dense networks generating life. The Universe is a system of deployment of informational networks to generate elements of all sizes and all structures.

We find here with the substrate of the control agents forming a global informational network the role given by the dark matter network defined by the physicists.

The molecules will not be formed by a simple chance meeting between agent atoms that are close in space, but by the tendency to a composition of agents that is given by the organizational law exerted by the constructive communications between the sets of informational envelopes of agent atoms. The consideration of random formations in a large set of elements leads to sophisticated mathematical models which are, unfortunately, not applicable here. We posit that there is a general organizational tendency that operates everywhere in the Universe and at all its scales and that is represented by a network of informational communications between all agents and informational multi-agent systems. The Universe is not a planned organization from the top down nor from the bottom up, but it is a self-adaptive organization exercised at all levels in sets of elements that communicate to realize aggregated systems that can develop more complex organization. We define the organizational law as the general constructive tendency of our universe, using the links between the innumerable informational envelopes of the elements forming the major network of information exchanges.

4.2. Formation of simple molecules of helium hydride and dihydrogen

We consider the Universe at the stage of development where it is constituted of a considerable set of informational agents and atomic agents constituting very dense groups in space. The control system uses the morphological field of each group and then of each component to continue the formation of the atomic agents. But there will be many atomic agents in the vicinity which will put themselves in strong communication and this will lead to the formation of molecules.

The expression of the organizational law in atomically dense environments

We are in the state of the Universe system where there are a lot of atoms created, especially hydrogen and helium atoms. Many of these atoms are in dense groups and they have informational envelopes that will communicate and form informational layers. A global informational current will be established in the groups that will allow all these elements to carry out coordinations, to form dense aggregates using these atomic agents. The importance of the density of the agents is significant, and it is this that will enable them to activate their rules of association. We then enter a state of the Universe system where the constitution of molecules will be considered as being the new major action of organization, because it will be carried out in most of the groups of atomic agents.

The atomic agents will aggregate, and we will therefore have the very important creation of molecular agents, which will form a system in which molecular agents will rub shoulders with atomic agents and informational agents. The system becomes very complex.

In this context, the process of creating a helium hydride molecular agent, which researchers consider to have been one of the first to form, is as follows:

1) There is strong communicative activity between a helium atom agent and a hydrogen atom agent, the two agents being spatially close in a dense activity domain.

2) The association rules of these two agents are activated via the incentives of the control agents.

3) There is realization of the connection of the two agent atoms by union of their envelopes and constructive distribution of the atomic elements.

4) There is a transformation of the informational envelopes of the two agent atoms to generate a new informational envelope of unification representing their union and allowing them to stabilize it. The envelopes of

the two atoms remain internally dependent on the new informational envelope of the created element.

5) The new agent has a main rule to guarantee its stable permanence and continue the general organization, seeking to associate with agents with a similar structure.

There will thus be the formation of many helium hydride-type molecular agents, which the organizational law will encourage.

The formation of a dihydrogen-type molecular agent can also be realized, because it is a question of creating a stable structure by realizing a covalent bond, putting in common the two electrons of two hydrogen atom agents. This is a rule that is not major in the informational network, and this realization will therefore be local and will not engage all the agent atoms. Let us assume that two hydrogen atomic agents are very close to each other and that their informational envelopes enter in communication. These informational envelopes activate their rule to perform an aggregation in the current context and they have no other incentives. The algorithmic process of creating a dihydrogen molecule agent is then as follows:

1) Communicative activity of the two very close agents of hydrogen atoms.

2) Dominant activation of the association rules of the two hydrogen agents in a context appreciated by them as dense and inciting control agents.

3) Realization of the simple covalent bond between the two agents, each one bringing its electron agent, the two electron agents being then put in common, forming a common information field.

4) Transformation of the structural union of the two hydrogen agents by constitution of the new dihydrogen agent, deploying its new informational envelope from the envelopes of the agents constituting it.

5) The new agent has new rules of action defined from the rules of the hydrogen agents and it has an informational envelope that ensures its stability and its local role, which is simply to maintain itself in action with the nearby molecular agents.

This molecular ion-type agent seems to be the first ion to be created in the Universe because the union of two agents of the same type can be done only among a very dense set (Gusten 1944). This is achieved because the tendency to unite agents is systematic and they will communicate to achieve it. Some unions will be stable, while others will not hold and there will be rupture and distancing of the agents. There will thus be, with the very strong density of the hydrogen atom agents, the formation of many agents of molecules of hydrogen and dihydrogen. There will also be the formation of molecules of hydride of helium, uniting a hydrogen agent and a helium agent. The formed molecules, which have stability in the sense of the organizational law, will not fracture and they will be able even to commit to changing their structures by becoming other molecules with more atom agents.

There will thus be, in the space that is dense in atom agents, creation of very numerous molecules that the organizational principle will make constitute themselves and gather, and there will thus be creation of molecular aggregates which will be the first elements of matter whose physical mass will exist. We can then specify according to which principle of organization all that is carried out.

The multiscale incentive exercised by organizational law

The innumerable generating elements that are the informational agents are incited to action by the informational substratum of control constituted by the informational network that assures an incentive control from a continuous morphological analysis of the organization of the multiple fields, generating incentives to aggregation and transformation. The Universe is strictly informational; it is a set of active informational fields, some of which are stable and form physical matter. The incentive control operated on the informational layers determines the state of the elements according to their morphology, and then the state of the fields of elements where their actions present similarity, thus releasing aggregates having uniformity to generate their organizing modifications. This control encourages the aggregates to unify constructively, to become more complex so that new elements are generated by transformative unions, which increases the structural complexity of the system (Cardon 2005). At this level of development of the Universe, the morphological and semantic control is realized everywhere and at all scales, which is the function strongly incited by the control so that new stable and strongly structured elements may be generated.

We must consider that the formation of a molecule under incitement of the substrate of control is the fact of the action of the informational envelopes of the agent atoms that unify to define the envelope of the molecule, but this is not the most important thing. The formation of this molecule, an element with stability and original structural complexity, will be apprehended by many other atomic agents through the information given by the action of the organizational law to realize, where it is possible, this same molecular type. Molecular agents are not formed in certain places by chance, but by the application of the organizational law which analyzes the morphology of the spatial domains, inciting them to make new creations similar to the first creations of molecules which were easily realized and increased the complexity of the organization. A local construction action thus generates incentives to perform this type of construction rather globally, by having the atomic agents form the existing molecules under strong incentives from the substrate's controlling agents who modify the importance of their rules when there are many specific successes. This is how, with this multiscale organizational principle with incentives from the controlling agents, the Universe will be able to organize itself in multiple structures having similarity, not forming a more or less chaotic whole produced by chance.

We thus posit that the Universe is generated with an evolutionary incentive control coming from its controlling substratum, a control which is realized at all scales in a continuous way in its deployment. Its evolutionary existence is based on this organizational law which is applied by the manipulation of the informational fields created by its multiscale self-controlling space. Therefore, we must assume that there is self-controlling space that envelops all the generated elements and there are those innumerable fields where organization and evolution unfold.

We will now see how this organizational law led to the generation of the material elements of the Universe such as stars and planets in the immense spatial domains formed by the galaxies.

5

The Formation of Physical Elements in the Agent Approach

We consider the state of the Universe where innumerable particulate domains have been generated containing dense clouds of informational agents, atomic agents and molecular agents in an immense space. The organizational law is applied to organize these expanding sets, and it will be applied very strongly by constituting great aggregates endowed with a proper capacity to become complex. Without this organizational law, space would have continued to expand with its informational agents and the clouds would have become a fog of informational agents and atomic agents that would have simply produced molecules in many places. For in order to form new aggregates with a good structure having permanence and leading to the notion of mass, an aggregating and organizing force is needed, and the organizational law will be this incentive force. The organizational law will exert its two major influences everywhere in the informational space:

1) It will continue the expansion of space by deploying informational envelopes around the very large aggregates and the informational space at the periphery of the general spatial domain.

2) It will combine the elements formed of the atomic and molecular aggregates to make them more important, more massive in order to continue the deepening of their organizations by the formation of new molecular elements. It will give this very strong incentive in the very big informational envelopes of the big sets of elements which will then be in internal organization.

5.1. The notion of aggregate mass

Many aggregates of molecular agents will form with mass, the notion of mass being the following, to which we will attach that of gravity.

The notion of mass in an informational space and the generation of gravity

In a very dense set of molecular agents, the action of the organizational law will lead to the formation of multiple aggregates with these agents. These aggregates will take on a structure and will be endowed with their own informational envelope guaranteeing structural permanence and the whole will also have an envelope. Each molecular aggregate will take informational energy from the speed of expansion of the Universe by taking energy from the substrate and it will use this informational energy to fix its structuring in time, to link the molecular agents that compose it and to make them operate with good communications. It will have a specific informational envelope to keep its structure and to attract other agents, atoms and molecules and other aggregates of smaller volume. All these aggregates will therefore have what we will call their **mass**, which will be the informational energy used to maintain their structure over time. This mass will be produced by the action of the informational envelope linking all the internal elements to the aggregate with a global force exerted for the attraction of other elements located in its vicinity when they are of smaller size, which will be the effect of what will be called the **force of gravity** of this element endowed with mass. Here is constituted the notion of gravitational force through the effect of attraction exerted by the unifying informational envelope of the elements of the aggregate endowed with mass which can **move in the space** by using energy.

We can therefore consider that these innumerable aggregates which are simultaneously constituted in the clouds of elements that make up the space of the Universe will have centrifugal forces to preserve their structures thanks to the organizing action of their informational envelopes, thus constituting pockets of matter. These aggregates will be able to move in this space, using energy, their informational envelopes pushing for a moment the informational envelopes of their components. The structure of the Universe is thus in modification; there will be very many material aggregates which will be formed and which will attract each other and will move physically.

We can consider that these material aggregates are **material agents**, with the following properties.

Material agent representing an aggregate with mass

– Precision of the internal structure of the aggregate, with the quantity and morphology of the agents that compose it and the informational fields that link them all. Analysis of its specific morphology.

– Precision of its mass, related to the quantity of elements and the type of elements that compose it.

– Precision of its informational envelope and action of attraction for the external elements smaller than it with possible integration of new elements.

– Action of attraction by an element of greater importance and possible union or absorption.

– Situation of equilibrium between aggregates that weakly attract each other according to their distance and keep their positions.

– Informational communication with other material agents and physical movement.

The formation process of the aggregates endowed with mass is indeed an action of the organizational law. There is a condensation of hydrogen and helium atomic agents and informational agents to generate aggregates. There is formation of an attractive envelope around each aggregate and formation of very strong communication links among the elements of the aggregate to constitute its current structure. There is generation of new atom agents inside these aggregates through unions and transformations of the initial atoms. Each important aggregate is a generator of new atoms and new molecules by development of the internal multi-agent system. The informational envelope of each aggregate attracts new atom agents and molecules to increase itself and this is its attractive action. This attractive force will be all the more important that the aggregates will be extensive with big and very powerful informational envelopes. The informational envelopes of the aggregates having mass are separated by domains of space, still containing atom agents and informational agents that are attracted, and therefore these domains empty, thus constituting what we will call the spatial vacuum and which is a reduced informational domain.

The process forming the aggregates with mass can be described.

Formation process of aggregates with mass

1) The state of the domain is characterized by a very large space where there is a quantity of clouds of atoms and simple molecules in movement due to the action of the permanent organization process.

2) Organizational law action: incentives for clouds of simple atoms and molecules to aggregate to form much denser structures using information incentives from substrate control agents.

3) Formation of dense aggregates by condensation of the elements of the clouds of elements and union of the internal elements by strong informational communications.

4) Formation of multiple very dense structured aggregates composed of atom and molecule agents, whose structure having permanence is considered to have a mass that is given by the informational envelope using the internal informational network between the elements forming its structure.

5) Development activity in these aggregates and formation of new and more complex atomic agents inside due to the organizational force engaging to generate new elements from these very neighboring active elements. The new atoms will have more than two electrons and there will also be formation of new molecules with informational envelopes defining their constant volumes and being able to move in the space.

The aggregates having mass will develop in a very strong way in the immense clouds of atom and molecule agents, because the organizational law encourages the pursuit of an ever-more complex development in the new aggregates, by general application of its rule of organizational similarity. All these aggregates will give birth to new atoms and new molecules in their increasingly complex structures which allow them to amplify the complexity.

The aggregates formed by atom and molecule agents have an informational envelope leading them to assemble with other very close aggregates to form larger aggregates. This is done taking into account the density of the aggregates in the huge clouds and will not lead to a single huge aggregate, due to the continuation of the expansion of the Universe and the initial fact that a multitude of aggregates are formed simultaneously, separated by empty informational space. There is the beginning of the development of new atoms and molecules in these aggregates in very great organizational increase. We are within the framework of the classical laws of physics and chemistry regarding the generation of atoms and molecules in a complex set with very strong dynamic organization.

The action of the organizational law on swarms of aggregates in an immense space is thus to provoke the generation of denser and more complex elements in each domain considered; it amplifies its action so that this is realized everywhere in the expanding universe system. It is really a law which takes strongly into account the organization of the domains at all levels to amplify their organization and generate new elements. This has been very well studied in physics and chemistry.

5.2. The formation of stars and galaxies by the general action of the organizational law

We are in a stage where the Universe is constituted by innumerable clouds of aggregates constituted by atom agents of the hydrogen- and helium-types and with molecule agents. There are aggregates with mass, which are in development. There are always partial condensations of aggregates and there will be the formation of massive elements. This stage is going to be the new form of development of the Universe, after the one of the informational agents and the atomic agents in intense multiplication. It is different from the previous stage, and the organizational law will change its incentive action. There will be considerable communication between the informational envelopes of the aggregates to form very big aggregates that will unify.

The specific action of the organizational law to transform the clouds of atomic agents and molecules

In the very active and very creative clouds of atom-agent-type elements, the action of the organizational law by the control agents will engage these atom agents to form aggregates and aggregates of aggregates, by diffusing to them a lot of informational energy. There is thus a morphological action at the maximum level of the Universe, to pass from the formation of clouds of aggregates to considerable sets unifying quantities of aggregates. The organizational law will considerably amplify the volume of the sets of aggregates to go towards something much denser and more complex; it will strongly incite these aggregates to condense to form very massive wholes having considerable volumes, and it will make them aggregate in a continuous way. There, the structure of the Universe, which was formed of innumerable clouds of agents moving in the vacuum, will completely change to see the formation of sets of elements of considerable size and mass, separated by practically empty space, and which will form the first generation of stars.

In these massive aggregates where all the elements are in very strong communication, there will be the formation of new chemical elements, atom agents of carbon, oxygen, nitrogen, fluorine; all the elements that have stability will be preserved and those that are unstable will disappear. There will be condensation of the clouds of molecular and atomic agents in multiple aggregates and aggregations of aggregates, with informational envelopes ensuring their temporary unification. The organizational law commits to the systematic formation of aggregations of aggregates so that the complexity of the atomic agents increases systematically. This will be the formation of the first generation of stars, which are considerable unions of atomic and molecular aggregates in absolutely intense organizational formation. These enormous stars aggregating quantities of atomic and molecular aggregates having mass will have a considerable internal activity, leading to the formation of many new atom and molecule agents, corresponding to a multitude of new chemical elements. There is an intense action of the dynamic organization of these elements in constitution, which the organizational law commits to produce to go towards much more important structures than those of the stage of formation of the innumerable clouds of atom and molecule agents.

The transforming action of general scope of the organizational law

The organizational law will act at the morphological level of the whole universe and will use the very dense clouds of atomic agents for the formation of considerable aggregates. In these new aggregates, there will be the formation of new atoms and new molecules which will be in intense interactions. It incites all the aggregates to aggregate by using their informational envelopes to form new meta-aggregates, with intensity. There is thus formation of new atom agents much more complex than hydrogen and helium but it engages with intensity the elements of the Universe to form very dense sets of elements, very massive, in considerable organizational state but which are not very stable. These will be the first generation of stars.

However, these enormous first-generation stars with considerable internal development will be too unstable, too large in volume with too many internal energetic deformations at multiple levels and they will explode by expelling into space multiple chemical elements. There will be a contradiction between two morphologically opposed structural tendencies of the organizational law: the tendency to complexify the elements of the minimal level to form new atoms and new molecules and the morphologically global tendency for the Universe to have elements of considerable size in its immense space.

The formation of a star of first generation is a process where the organizational law will morphologically operate at the level of the extent of the Universe, uniting considerable sets of aggregates. It will engage to realize the maximum internal organizational development of these considerable masses of active elements so that they may generate new chemical elements.

Formation process of a first-generation star

1) Intense activity of clouds of agents, atoms and molecules forming active aggregates.

2) Very strong connections between the organizational envelopes of the aggregates of agents to unify in increasingly important groups with important acquisition of informational energy of the substrate.

3) Formation of a considerable cloud of aggregates of atomic agents with very active domains and others that are a little less active, and continuous attraction of all the neighboring aggregates.

4) Unification of the informational envelope of the groups of aggregates, constituting their informational envelope with distinction from the contextual environment.

5) Action of the global envelope of the aggregates to very strongly pursue the condensation of the aggregates between them which form thus a very very dynamic field.

6) Condensation of innumerable aggregates of the elements globally enveloped forming very strongly unified sets, with envelopes which will exert an internal force of intense pursuit of the generations of new atomic and molecular elements and an external force of attraction, this force depending on the size of the aggregated set. The first-generation star is in formation.

7) Intense internal activity of the enormous first-generation star, continuous formation of internal aggregates with new atoms and new molecules. The organizational law intensifies these new formations to increase the complexity of the atomic elements of the Universe.

8) The whole lacks overall organizational coherence because too many internal aggregates form and clash with each other.

9) General instability of the first-generation star compared to the informational whole of the Universe, because it continuously develops in its internal space too many hyperactive, distinct and opposed elements.

A lot of new components will be generated in these stars in their internal activities in intense modification. This will be the creation of all the important atoms produced in complex masses, but such a hyper-generative set cannot be stable and the organizational law, operating at the general level of the Universe, will make these stars explode continuously. It will be an action of rupture of the forces of proximity between the masses which form them. There will then be the formation of second-generation stars in galaxies, stars which will have spatial and temporal stability.

The innumerable aggregates of atoms and molecules expelled into the informational vacuum by the explosion of the first-generation stars will undergo again the organizational law of formation of large aggregations. These masses will aggregate and constitute in the duration the stars of second generation, which will have much less important sizes and most importantly much more stability, and will possess the right atomic elements to develop further. These stages have been studied in great detail by physicists, because the second stage of star formation is the one in which our universe finds itself today.

The explosion of the first-generation stars will be the next phenomenon. The intense action of the organizational law to produce the enormous stars of the first generation focuses on the formation of considerable masses in relation to the volume of the Universe and also the search for the production of new atomic elements by the action at the level of the atomic aggregates. These two focuses engage an intense activity on the aggregates to make them produce new atomic agents, to amplify them, to make them confront each other. There will be the formation of elements of considerable size which will be the first generation of stars, but the internal energy spent is too important; there will be a very strong internal chaos and physical explosion of these elements. The organizational law will then immediately act in an intense way on another scale, still both globally and locally, by inciting the physical elements formed by the explosions to aggregate to constitute large physical elements of more reasonable sizes, which will be the second generation of stars.

The stars of the second generation will be stars moved by strong rotation operated by their informational envelope and thus by their gravity to maintain in a state of general stability their structures with their internal elements of dense and very active masses.

There will be groups of second-generation stars formed by the physical constituents of the first-generation stars, and these will be the galaxies. The new galaxies formed by groups of second-generation stars will be groups where the first-generation stars fragmented to give much smaller, denser and more stable stars for the formation of new elements internally, and separated by informational space. There will be galaxies formed with many, many stars with physical stability, which will be grouped in informational envelopes ensuring permanence to their structures. There have been two different actions of the organizational law. Initially, the law generated considerable domains of dense elements so that innumerable new atomic and molecular elements could be formed by maximum local actions, and then these great domains of elements became energetically unstable and chaotic due to their incessant internal energetic activities; there were physical explosions, and the organizational law committed the elements coming from the explosions to form the stars of second generation, more stable, with the formation of their planets.

Formation process of a second-generation star

1) Explosion of an unstable first-generation star and formation of innumerable parts and free elements in the information space.

2) Action of the organizational law to condense the parts forming clouds, and formation of new large structured aggregates with good local informational envelopes, which will form much smaller and more stable stars by condensing their elements.

3) Generation of multiple, much more stable second-generation stars with envelopes limiting them in an environment still containing clusters of atoms

and molecules. These sets of stars will form the new galaxies caught in the global informational envelopes.

4) Action of the informational envelope of each star to keep its structure by carrying out its rotation on itself in order to amplify the condensation of its elements and obtain its final structure. This action of the envelope will cause a movement of rotation of the clouds of elements which are in its periphery.

5) Action of the organizational law to condense the clouds of aggregates in movement around the star: we go towards the formation of planets.

6) Generation of planets around the stars, by regrouping the aggregates of the peripheral domains and rotation of these planets around their stars to preserve the structure of the whole while respecting gravity.

So, we have the formation of planets around new stars.

Formation process of the planets of second-generation stars

1) Activity of the organizational law on the clouds of atomic, molecular and material aggregates in the peripheries of the new stars to realize strong condensations of elements.

2) Formation of planets by condensation of these clouds of elements with the creation of a local informational envelope ensuring the development of their structure.

3) Action of the organizational law with a stabilized attraction generating the rotations of these planets around the stars and on themselves.

4) Development of organizational activity on planets internally and on the surface if their organizational state allows it.

We can therefore consider that the development phase of the Universe was a phase of intense activity at all levels, from the initial level of the informational components generated by the generative component, to the level of the formation of the atomic and molecular agents which generated multiple aggregates leading to aggregates of considerable volume producing the first generation of stars. Then, these first-generation stars, too unstable due to their internal energetic development, exploded and produced the second generation of stars, which had much more stability and located their planets in their peripheries. On some of these planets, an organization of the living could be deployed. We must therefore consider that the Universe system is a system produced to continuously develop by producing and regulating the generations of elements which are constituted, by making the elements, which have structures that are stable in space and time, exist. The principle of the organizational law is to make the organizations continue without planning a final state to be reached, because it acts in its present through local and global analyses.

The general principle of organizational law

The organizational law uses its controlling agents to give incentive information and informational energy to groups of changing agents. It acts on some elements, and if the result is good, it transfers this incentive to multiple other elements. This is what produced the huge swarms of atomic agents from the basic informational agents, and then the organizational law operated at the level of the space of the Universe and this led to the production of innumerable atomic aggregates which assembled very strongly to produce the first-generation stars. There was no global regulation to limit the formation of these stars which were too large and developed far too much internal energy activity. The organizational law uses controls on informational envelopes and generalizes its incentive action without planning the final effect, which is not predictively calculated by the control, to leave autonomy to the Universe in organization. It is a law of action of organization making the current state evolve towards a following state by the development of modifications.

We can insist on the truly organizational role of the stars in the constitution of a universe having structural permanence.

The organizational role of the stars

– The stars of the first generation and especially of the second generation will generate new atoms and new molecules, by organizing considerable and very compact masses in which the internal morphological movements of transformation will be very important. The stars of the second generation are the fundamental elements of the Universe because their structures have very good temporal permanence and considerable energy. They will send great flows of photons into the surrounding space as well as clouds of new atoms and molecules, which will influence the organizations on their planets.

And we are going to consider that stars and planets are systems that have an **operational closure**, that is, a functional and informational envelope allowing the continuation of their developments in a very deep way in their own organizations.

We have therefore posited that the activity of the organizational law on a set of aggregates in very strong communication will lead to the formation of first-generation stars, with considerable internal activity proportional to their masses, and then second-generation stars with planets having stability. We can summarize the actions of the organizational law that has been expressed on all the informational envelopes of the generated elements, since we have posited that the Universe was founded on a totally informational substratum represented by the informational fields of communications and actions.

The actions of the organizational law

1) We have posited that the construction of the Universe was based on a singular element generating considerable informational energy that produced a quantity of informational agents, and these generated either structural components that simply produced informational space, or active components, and this in a systematic and expansive way. This was the first phase of action of the organizational law: to continuously activate the available informational agents to generate informational space by reduced agents or to activate active agents to produce aggregates. Then, when the active agents were very dense in the created informational space, there was the action of the informational tendency exerted by the informational substratum so that they activated their quantum properties. Thus were formed the first clouds of quantum agents, which formed immense clouds.

2) The informational envelopes of these basic quantum agents caused them to aggregate to form the basic atomic agents, informational agents.

3) The informational envelopes of these atomic agents coordinated to produce sets of aggregates that formed the first atomic agents of hydrogen and helium in immense clouds separated by vacuum, and some molecules.

4) Then, the informational envelopes of these atomic agents coordinated to unite aggregates of agents and form the first structured elements with stability and mass. There were then many elements forming multiple domains of aggregates in swarms as well as isolated atomic agents.

5) All these aggregates were coordinated by very strong incentive of the control substrate in their informational envelopes so that they formed enormous sets of elements which became the first generation of stars.

6) These initial stars were very active at the internal organizational level and generated many new atomic agents and new molecular agents. The organizational law greatly increased the complexity of the generated elements by increasing the categories.

7) But these first-generation stars were too dynamic internally, with activities constituting non-uniform domains, and they exploded, splitting to form multiple clouds of massive and dynamic elements. Action of the organizational law provoked condensation into much more stable elements, very numerous and well distributed in the space vacuum, which became the second generation of stars.

8) The clouds of free aggregates around these second-generation stars aggregated due to the strong incentive of the organizational law and gravity to form planets.

9) And then, there will be development of organizational activity on planets where the conditions will be favorable, being well located in the regular space-time.

This shows that what is fundamental in the Universe and in its constituted elements are the informational envelopes that realize the communications by representing the incentives of the organizational law emitted by the substratum. These envelopes allow the generation of stable elements in temporality. We are in a universe of a truly informational nature, designed to organize itself by creating active organizations at all possible scales, and without any higher direction.

We can still insist on the continuous action of the organizational law which is the action of the substrate producing the space-time and the elements of the Universe by emergence. The organizational law produces the permanent action of the fundamental informational substratum which apprehends the elements and the emerging domains of the Universe at all scales and incites elements and domains to realize aggregations and to increase the complexity of their elements, by sending information in their informational envelopes, through the messages and the informational energy emitted by the informational substratum and inciting to continuous organizational actions.

The creation of material aggregates by the incentive of the organizational law operating through informational fields has resulted in structures that will be difficult to move or destroy, and in these cases, it will be necessary to use significant local energy to modify the organizations. So, there exists both informational energy that is underlying everywhere and physical energy for structural changes, much used by humans, which is totally localized and can very often be significant.

Two forms of energy in the Universe

There is the informational energy that comes from the substrate and represents the communicative actions of the informational fields using the universal web of informational links, and there is the local energy of any activity and of any provoked modifying action, which is the physical work necessary to realize movements of physical elements or modifications of the structures of these elements. There is non-symmetry between the relatively weak force used to generate the aggregates with informational energy and the important force necessary to modify these aggregates constituted by permanent informational links realizing their atomic and molecular networks.

We can posit that the intense explosion of stars can produce **black holes** where the material emergence of the Universe has disappeared and where there is only a cell of access to the informational substrate and where absolutely nothing is accessible to us. Can this be seen as a local end of a local part of the Universe, and would the explosion of all the stars of a galaxy produce a considerable black hole which would be an access point to the substrate and from which another universe could be produced?

5.3. The informational program for the design of the universal system

The creation of the Universe was done in three main stages. First, there was the construction of the generating component, which was realized in the Universe of the builders, then there was the putting into activity of this component, then the realization of its generations of components, by supplying it with the necessary energy, and it truly is energy coming from the world where the generating component was generated. We describe the stages of the generation of our universe considering that it was realized as an informational program.

Step 1 – formation of the generating component in the generating system

The production of the generating component amounts to producing an information-type system, of general computer design but operating on very different basic elements from those of computers, and which has three structures:

1) A structure generating informational components that will be informational fields and considered as informational agents.

2) A structure that continuously generates in the operation of the cells made up of sets of informational links enveloping the components generated by connecting them and that will form the space available for the movements of the components.

3) A knowledge-generating structure in informational agents which produces and sends through the network of links created between components' organizational information to these components and to the aggregates of components so that they organize themselves.

This design is of the computer-type, but the elements constituting the structure of the generating component are absolutely not electrical impulses localized on electrical or electromagnetic networks. They are morphologies of informational fields, fields of a type similar to the quantum fields and whose very well-structured whole will make the program of the generating system. The designers knew how to program sets of information fields whose morphology will make the programs and their subprograms. There will be morphological sets representing instructions, commands, memory elements, processes. There will be sub-programs and functional links between them, and there will be an informational envelope defining the current effective action. This component will thus have the structure of a system executing a program and modifying itself by its activity. All the executed commands will launch small programs, whose morphological form will characterize their characters, which are to produce new informational components or cells of the substrate.

This type of general program is perfectly conceivable in computer science, in particular through the development of autonomous multi-agent systems controlled by morphological agents. It is the representation of instructions by informational fields that poses a problem today.

Step 2 – activation of the generating component

The activation of the generating component is the launching of the program it represents. It has energy supplied to it to activate all its local informational fields. This energy is informational, giving the morphological structures constituting the structural elements of the program the possibility to activate and generate active components. It thus has a computer program structure, with an initial program that is launched and that will activate all its necessary sub-programs.

There will be two main subroutines activated by the generating component. There will be the subprogram of generation of a cellular element constituting the substrate and containing the informational links which will be, in the emergence of the Universe, spatial vacuum. It is the creation of a cell whose informational activity consists of producing communications expressing the characters of the organizational law. This cell will thus be able to observe the activity of all the active informational components by simply analyzing their informational envelope and then to send them organizational action messages. This cell will contain all the links between the components, links that will be permanent wire fields. It will be in communication with the Universe of constructors who will be able to transmit informational energy to it, which it will give to the active components.

For the establishment of these links, we must assume that the whole of the active components will form an immense hypergraph enabling there to be an address for each component. We can then consider that the whole of the cells of the substrate form a considerable **organizational memory** locating in fact all the elements of the Universe system. Each of these cells has a small local action program which analyzes the states of the informational components of its domain to commit them to aggregate or modify themselves, or to move. The whole of these cells thus forms the active informational action network of the substrate of our universe.

There will be the subroutine of the generating component launching the creation of an active informational component. This is the launch of an informational action agent program with specific informational subroutines so that it can produce other components and substrate cells. The created element is placed in a substrate cell in the neighborhood of the generating component. Each generated element will be an action element, with informational energy transmitted to it by the generating component, which it will then use in its generations of components or cells that will have energy coming directly from the substrate and the creators' universe. It will be able to activate its sub-programs which will be the elements of connection with the other components and simple messages requesting aggregations.

There will be the control program of the generating component and then of the substrate elements, which will control the frequency of all these activations and generations, and this in the right order to satisfy a good distribution of the components in the created space by providing them with good complementary quantum characters.

Step 3 – the formation of quantum, molecular and mass elements

The action of the active components in the cells of the substrate will initially be of intense generation, with a lot of informational space generated and the formation of swarms of active components generated in reproduction by the components themselves. The creation will be in a good order set by the control constraints of the substrate. There will be many generations of space cells that will unite to form an immense expanding memory space in which the active components will form swarms. When these swarms are numerous, the organizational law will change its incentive communications to the components to solicit the joining of components to generate quantum elements. There will then be communications so that the components recognize each other as complementary at the level of their fermion-like quantum elements, and the first atoms will form. This will impose their spatial displacement for the reunion.

The spatial displacement of a component is strictly informational. It is a command from the substrate that switches off a component, absorbs it for a moment in its field, crosses links and then reactivates it in the proximity of the complementary component, at another place in space. The substrate thus serves as a transfer link between the structures of informational fields constituting the components to move them from a place to another. There is indeed in the hypergraph constituting the network of the links of the substrate the spatial addresses of the components, which is the case for any dynamic memory.

The atomic elements have structures that must be permanent; they are structural elements of the emergence of the informational space constituted by substrate cells and components. An atomic element is no longer strictly in the informational domain: its informational fields are, but its informational envelope is in the space of the Universe where the physical rules for any displacement are exercised, with expenditure of energy and notion of the duration of the transfer.

The formation of the molecular elements will be carried out in a different way than that of the atomic elements. The displacement of the atom elements to generate molecules, after the incitements given by the communications of the substrate, will be done by attraction and will require the expenditure of a little energy. There will be exchange of communications between the atoms, then association to form a molecule and there, the displacement will be done in the space of the Universe at the physical level, with expenditure of energy, of rather photonic-type. There will be physical transfer, with avoidance of collisions by the work of the informational envelopes of the atomic elements on the trajectory.

Then, we will arrive at the sets of atoms and molecules forming aggregates endowed with mass, with informational envelope defining the notion of gravity. The organizational law will amplify the action of reproduction of informational components in an immense space created, and swarms will be formed. These clouds will generate aggregates of aggregates, and after a long time, there will be the formation of first-generation stars. The formation of these stars shows that the organizational law acts on the current state of aggregates and does not globally evaluate the whole of its realizations. For these stars will be unstable at the informational level and will explode. The substrate operating the controls is indeed a local control network that communicates with the informational envelopes of elements that are not of considerable size. There will be in the immense space invaded by clouds of elements the reconstruction of new more stable stars, with their planets. We will see in the second part the program that has produced life on the Earth.

The realization of the program generating a universe

The program of the realization of a universe is conceivable at the level of an information system self-adapting to itself and using this to deploy itself, in particular by using multi-agent systems with an autonomous and continuous morphological control. But the actual realization must be done under the quantum level, which is for us difficult to realize. But it will be possible to use quantum computers to activate quantum elements in artificial vacuum created by particle gas pedals, with the use of photon fields to generate some creations of informational components and the formation of new atoms.

We are now going to see how evolutionary life could have been generated on our planet Earth.

PART 2

Life Produced by the Organizational Law

Introduction to Part 2

We are now considering the Earth, within its delimited space, becoming stable and therefore a morphologically very different domain from the organizations taking place in the clouds of space. The evolution will be on a different scale, which will impose modifications in the application of the organizational law, as we will see.

Any understanding of life on Earth cannot be based on a simply functional approach, since it is widely considered since the 20th century that there is a very large organization that unifies species. The problem is to determine the cause of this considerable evolutionary organization that has generated countless coherent species. There are researchers who consider that the self-organization of life is autopoietic. A system is said to be autopoietic when it is an autonomous organization that carries out its regulation and evolution by regenerating its components and their relations in a continuous way, being in fact a network of elements undergoing continuous restructuration by its own functioning (Varela 1989). Our informational model will be based on this approach, and we will try to precisely define the actions of the organizational law that generated all life and realized its evolution.

It is a question of considering living organisms as having a highly communicative functional structure and seeing where they come from in the evolutionary organization. For this, the scientific approach is based today on DNA, which is a structure made of molecular conformations from which a plan totally defines a living organism during its creation. So, we adopt an ascending structure, going from the DNA to the organism conceived, then to the species and to the transformations producing other organisms and other species.

But the question that we ask and that has led to the refinement of our model of life is the following:

– Why and how was it possible to generate life in the different geographical areas of the Earth and why were there so many generations of species with permanence and very good behavioral abilities?

To answer this question, we will have to start with the notion of organizational fields on the planet Earth and consider that everything that exists there comes from elements that are multidimensional informational fields, that all molecules and molecular aggregates are elements that structure their informational envelopes allowing them to communicate in order to organize themselves continuously at much larger scales.

We will posit that a living organism comes from a progenitor organism that has a strong informational envelope around it and obviously around its DNA, which then plans the generation of the well-structured organs of the new organism to constitute it by communicating at the informational level with the living organisms of the geographical area where the progenitor is in action of creative production. We will see that this approach allows us to understand how and why the innumerable species were formed and who formed the living **Gaia** on Earth.

We will first present the different classical scientific approaches to life by asking constructive questions to introduce the presentation of our informational model. Then, we will present an informational approach and an agent approach showing how and why life was generated and how it evolves continuously, by introducing the very important notion of morphological pattern of informational agent sets.

6

The Characteristics of Scientific Theories of Life

We present different scientific theories of the living, theories that have strongly evolved over time and that are now scientifically very finely developed using very powerful concepts such as the role of DNA.

6.1. Evolution and selection: Charles Darwin's theory of gradual evolution and the biochemical approach

We must obviously understand that living organisms are not entities created to make life pleasant for humans, who have a place apart, which some people once posited. Living organisms share the same environments and often the same resources, some have disappeared and innumerable fossil traces can be found in ancient strata. Living organisms appear as the result of an evolution whose reason must be specified.

The minimal answer to the question of the reason for the existence of living beings consists of using the concept of possibility, by determining a domain of reality where this notion makes sense. We say that an event is possible if we can conceive it precisely and if we can measure its occurrence in a certain interval of time and space. The living thing that we see before us would thus be considered as being there because its creation was possible, because it is the result of a series of possible events that were realized, and this over very long periods of time. Life would then be simply an organizational transformation of itself in space and time through the production of many organisms.

Scientifically, living organisms were very early classified in a hierarchical manner. A class of living beings is a very general category where the general characters of the organisms have similar properties, the class being subdivided into orders and then into families, families where finally all the individuals are similar and differ only by a few traits. Aristotle had established a hierarchical classification of species according to their anatomical characteristics, obviously establishing that human was the highest species in the hierarchy. This idea of hierarchical order will continue with the development of comparative anatomy. In such hierarchies, we speak of classes, considering them as groups of organisms with distinctive anatomical features. The classes are distinct, the hierarchy consists of fitting them into each other according to the existence or non-existence of certain characters and comparing the qualities of specific characters. The way of seeing the classes consists of considering that there is a certain evolutionary tree which defines the characters of the successive transformations of certain classes into certain others, thus forming what is called a tree of evolution. Today this is the central problem for the classification of species, in the existence of this generative tree that is very finely specified.

The notion of evolution of species was formulated by Lamarck, before Darwin. Lamarck posited the existence of a tree structure to represent the evolution of species, in which some species derived from others, with more specific and more complicated characteristics. But Lamarck's thought was that the living being that evolved was the seat of a ***vital***, non-material, absolutely mysterious ***force***, a force that led the living being towards more and more complexity and perhaps towards perfection. Lamarck's evolutionary theory also stated that the advantages acquired by any organism during its life were transmitted to its descendants, which, according to him, justified an evolution with systematic improvement. Lamarck's theory was vitalist, thus positing the existence of an occult vital force leading living organisms towards their inevitable evolution, by allowing each organism to transmit its acquired advantages.

The notion of selective organic evolution was proposed by Charles Darwin in 1859 (Darwin 1859). The latter adopted a radical nominalist attitude, denying the notion of species as an *a priori* existing entity. According to Darwin, the species is the result of a selective evolution of organisms. It only exists if individuals are related, similar through the generation of descendants, and live together in the same geographical locations allowing them to reproduce. He will affirm, by his multiple observations and by taking into account the fact that each species gives life to more individuals than can survive, that the living species are the result of a systematic organic evolution operating simply by ***selection***.

Darwin also considered that each new organism that is born has some characteristics that are different from those of its progenitors, and that this is due to ***chance alone***, that the variation of characters due to birth is blind and that only natural selection that could give an advantage to some individuals that multiplied a lot, while others had practically no offspring. The advantaged individuals obviously multiplied more than the disadvantaged ones, and thus there was evolution by natural selection. He thus rejected both the fixist and vitalist theses.

According to the theory proposed by Darwin, a living organism descending from its progenitors varies little with respect to their characters, because it adopts and can only adopt the majority of their characteristics by organic inheritance. Then, it is subjected to the intense pressure of its environment, which will carry out the selection of the most suitable. The living being is then considered only as a transformation process which never stops as long as it can develop, as long as the environment allows it. Nothing is fixed, never, and especially not the species which can only evolve more or less slowly. According to him, the notion of species was no longer considered causal; it no longer existed *a priori*, but was a possible consequence of the numerous descendants of favored living beings. Some living forms are similar and constitute groups by their descent, and those ones we are able, with an aim of classification, to name "species", because these organisms are quite simply descended from the same ancestors and their environment allowed them to multiply and to live together, at the same time and in the same space that they share, and by crossing between them to establish the general kinship.

In Darwin's theory of natural selection, life is only a regular evolution based on two principles: continuous but weak mutation of reproducing organisms and natural selection by the strong pressure of the environment. The groups of individuals thus change weakly but systematically during reproduction and thus acquire specific characteristics, by the selection of those most capable of survival in competition with the environment. This is a theory of evolution which is called ***gradual evolution***.

The question at hand is obviously not the notion of evolution itself, which is scientifically proven by the discoveries of innumerable fossils of extinct species, but rather of understanding living beings in their entirety, in their situation on the planet Earth where geological upheavals have occurred. According to Darwin, the characteristics of living organisms change very little at each reproduction, so that the individuals are viable, but they change constantly, either a little or more strongly. This leads, through geographical isolation of some individuals, to groups becoming different during successive reproductions. Only groups with different characteristics and behaviors can coexist together in the same environment, sharing the resources of the environment. There is then a diffusion of varieties via the isolation of certain groups which become specific. Groups with very similar characteristics are in competition with others for space and resources; they can only compete with each other and therefore can either split up to survive or disappear locally leaving no descendants. According to this law of natural selection, order is created by selection between competing species, whose interrelationships constitute, so to speak, disorder.

The explanation of life, under the Darwinian approach, is a causal type of approach situated at the individual level: we start from an initial individual allowing us to conceive the reproduction of its characteristics, which was to be discovered at the time of Darwin, and we try to define all the intermediate stages of the evolution of living beings as a series of regular transformations in the production and selection of the different individuals. But the problem is that we do not observe the evolution of individuals forming the constitution of species and, above all, we do not know or find these intermediate forms linking an initial organism to its descendants that have evolved regularly. We note analogies and differences between individuals of different species, without having the chain constituting the intermediate forms. The validation of this theory is therefore problematic. Paleontologists have looked everywhere for intermediate forms, without really finding many. Darwin firmly believed in the notion of continuity and was convinced that nature could only create phenomena in a continuous manner, starting from a precise initial state, reaching an equally precise final state and passing through all the necessary and sufficient intermediate states. He did not conceive the notions of jumps, bifurcations or ruptures, which are modern concepts.

Darwin actually proposed two theories of evolution by natural selection, with two different frameworks of application. The first theory can be called the restricted theory of evolution (Denton 1993). It is about proving that species can derive from others in well-localized places, such as well-isolated island groups. This has been successfully proven in many cases. We have observed the gradual evolution of a species that has derived morphologically and gradually into many varieties, by geographical separation in a limited space, to finally end up with a different species with which crosses have become sterile.

But Darwin's second theory, which we can call a general theory, is that this mutation process should apply to all living organisms on the planet. Here, in this case, we change the scale completely. There is a lack of intermediate fossil species to support the thesis of gradual evolution. So, if there are no regular intermediate species between stable species, how can we explain the creation of so many very different and stable species everywhere on Earth?

Another approach to life, which is biochemical, is based on the consideration of the morphology of individuals: **ontogenesis**. Ontogenesis is the mechanism allowing the creation of an organism from a fertilized cell, by multiplications and cellular differentiations. Living organisms are then defined by the structuring and coordination of their organs.

A living organism is made up of very well-coordinated organs made up of differentiated cells, each of which is specific to the different organs. Each organ has its own cell-type or types, cells that regularly regenerate identically, except in pathological cases, which allows the organ to continuously regenerate so that it is always functional in the organism. It is indeed a multiscale organization.

It is possible to specify the typology of the living by defining the four necessary characters that all organisms qualified as living must possess (Danchin 1983):

1) A living organism has a metabolism: it continuously transforms certain elements of molecular origin into other active elements, to always be in a state to feed itself and to guarantee its temperature.

2) A living organism is compartmentalized: it is made up of parts that have an interior, an edge and an exterior. This is the case of the cell, which is the basic element of all organisms that are multicellular.

3) A living organism contains, in its structure, the way to build another organism which will be similar to it, which is the memorization of a plan or a scheme of construction for a new organism. This plan will be transmitted to every new organism created and this is the basis of the notion of heredity.

4) What is in memory to build another organism is exploited in the construction, developing the construction process. The organism can thus replicate itself in the form of organisms that it will build a certain number of times in a manner similar to itself, when the construction process is invariant. The construction process allows an identical copy, but it can have certain variations due to transcription defects: it thus allows an evolution of the living being by certain modifications of the generated organisms. These modifications are very weak during each reproduction.

According to these characteristics, some organisms cannot be considered as truly being living organisms. This is the case, for example, of viruses which are compartmentalized, which have a memorized construction plan, but which have neither metabolism nor their own construction process. The key point in this approach to the characteristics of living organisms is point 4. It is through the study of this point that the theory of living organisms will be played out, and we can ask ourselves how species that are very stable over time with very elaborate characteristics are formed, and whether organisms with new forms are formed by transcriptional defects in the construction processes. We will show that there is another approach.

6.2. The constitution of life, from DNA to developmental biology

The hereditary link between reproducing individuals is nowadays very clear and it is constituted by the genes, which are considered to carry all the information about the phenotype of individuals and which are directly transmitted in the reproduction process. Initially, these genes were considered as not modifiable during the life of the individual. The explanatory character of the theory of heredity was therefore essentially based on the unidirectional causal link between gene and phenotype. This was the fundamental principle that biology defined in the notion of heredity.

Research has found the basic element used to store hereditary information and the principle of reproduction of organisms based on this element. By transforming itself through molecular combinations, a living organism is then understood as a very complex sequence of biochemical processes leading in reproduction to the creation of a new organism.

It is considered that the fundamental building blocks of living beings are **proteins**. These chemical elements have specific forms and play the role of managers of the cells, because it is these chemical elements that regulate the activity of all the cells. They are the elements of activity of the cells, and we can consider that the proteins are similar to specialized devices carrying out very precise tasks. In order to organize themselves, so that the set of proteins forming each cell constitutes a coherent whole, it is necessary to have elements that carry information. These elements which carry information, which make it possible to plan the activities of proteins, are other molecules, the molecules of **nucleic acids**. They are informational elements, which activate or inhibit the activity of the proteins, and they are of two kinds: the **RNAs**, the ribonucleic acids, which directly transmit the information to the proteins, and the **DNAs**, the deoxyribonucleic acids, which are in the nucleus of the cells and which supervise the activity of the RNA, by activating them with injunctions given by information.

The central principle of molecular biology has well specified a relationship of order:

– DNAs control RNAs by creating them active: this is transcription.

– The RNAs allow the synthesis of proteins: this is translation.

– There is no returning feedback of proteins or RNAs on DNA.

DNA is not a simple molecule and its role is therefore considerable. It seems, consequently, by its structural complexity, that life is not at all an obvious problem. DNA is a polymer formed by **nucleotides**, and it has a spatial structure in the form of a double helix. Nucleotides are made up of a phosphate, a sugar and a nitrogenous base. These nitrogenous bases are four in number and are named by their prefix letters, A, T, C and G, thus forming four different nucleotides. A DNA molecule is thus an ordered repetition, of varying length, of these four nucleotides. The nucleotides are used to code, that is, they ensure the production of the 20 amino acids that make up all proteins, which are the major constituents of the cells of all living organisms.

DNA molecules can contain several million nucleotides, and thus form very long chains that can reach several centimeters. RNA is a single helical strand composed of four nucleotides similar to those used in DNA.

The DNA molecules therefore include specific message sequences, nucleotide sequences called **genes,** which will code for and form very specific amino acids in the cell. These genes are first decoded, that is, expressed in RNA, and the action of the RNAs will form a certain sequence of amino acids that will make up particular proteins in the cell. For proteins to be formed, a specific means is needed, a kind of protein mold. The protein synthesis apparatus is the cell's **ribosome.** It is a structure that is composed of about 50 molecules and that can make all the proteins that the DNA controls via RNA. All this constitutes a formidable synthesis apparatus, which can even manufacture the molecules, allowing it to duplicate itself.

So, there is an informational plane constituted by the genes of the DNA; there is the informational messenger which is the RNA, and there is the constitution of the corresponding elements, the right elements at the right time and at the right place, the proteins generated by the ribosome. This is how all the cells are formed, cells with the right chemical characteristics for each organ, of which they will be the basic elements.

All living organisms, from bacteria to mammals, share this same cellular system. In the fly and in human, as well as in all living organisms, we find certain identical genes to form the axes of future embryos. These two very different living organisms, the fly and human, therefore share genes, but do they have a common ancestor that would justify the theory of evolution?

The constitution of living organisms is a construction of great organizational complexity. A bacterium, this tiny elementary living organism, is composed of a hundred of billions of atoms and is organized with multiple proteins which allow it to exist, to feed and to multiply.

But the problems kept appearing. Scientists thought that humans shared 99.9% of their genes, which were identical and identified them almost completely. But it was discovered [published in *Nature*, November 2006] that humans differ from each other by 1% of their genes. They are therefore not totally similar genetically. The problem of the primacy of the genome has thus been complicated by this discovery.

It thus remains to be understood how well the genetic plan and the very powerful constitution process of the organs of multiple organisms could be constituted, so that a reproducer could produce a new very coherent and perfectly functional organism.

6.3. Genes and their expression: an open problem

We are going to see the scientific evolution of the genomic aspect in the reproduction of all life. The first living beings that were created on Earth came from water in large areas, where there was organization of water molecules and fertile minerals such as lavas, sands, metals and clays (Paccalet 1991). The fundamental elements of life are considered to be amino acids and nucleotides, which are composed of the oxygen, hydrogen, carbon and nitrogen that were present on Earth.

The problem is the creation of the first living things, bacteria, which are without a nucleus but have an envelope that makes them exist as specific elements with the power to duplicate themselves. Then, the bacteria were able to polymerize RNA to reproduce in a similar way, which generated countless RNA-genomes. Then, there was the creation of cells with the creation of DNA. But why was DNA created and how does the passage between DNA and the proteins that will form the differentiated cells of living organisms occur? How and why does DNA possess information that finally allows the generation of organisms that are different from their progenitors and well adapted to life?

Physically, genes are fractions of DNA and their action is *indirect*. Parts of DNA are transcribed into another molecule, RNA, which codes for the activation of different proteins. Genes contain the information that is used to define the phenotypic character of the newly created organism by importing and mixing the DNA of both the female and male progenitors.

Genes are grouped together in each cell in the form of long chromatic filaments called **chromosomes**. Chromosomes are therefore assemblies of genes. They carry, in sexual reproduction, half of the construction characters of the two progenitors, male and female, which are thus shared and available for the creation of the descendants and which are transmitted to them as their inheritance from the two progenitors. The number and type of chromosomes will characterize the species.

The main question about genes is then: how do genes operate and what global and local plans do they follow? There are several possible answers, and the hypothesis of determinism is important.

The deterministic rule

The assumption underlying the determinism of molecular biology is that gene expression is deterministic. There is a precise, non-random program that runs in an order imposed by the very structure of DNA and that produces the proteins of the specialized cells that form all the different organs of living organisms.

We can ask ourselves what is the notion of program that is used here, by specifying its very deterministic character. We consider that it is defined in such a way that to any precise situation corresponds a totally precise action that must take place: a precise cause produces a very precise effect.

DNA carries the genetic code of living organisms and therefore allows their reproduction. In order to create a living organism composed of multiple cells forming the different organs, there must be a manufacturing plan for all these cells, allowing their creation in the right order at the right time, and coming from the general plan used by the parent organism. This plan must be, in a way, deposited in the genes of the DNA molecules. The reproduction of living organisms is thus seen as a process using hereditary memory directly transmitted from sire to offspring by DNA. The whole standard theory of genetics is based on the notion of necessary and sufficient information used by the precise plan of the development of organisms.

This information contained in the genes will be used in a development plan, starting from a fertilized cell, to generate the billions of specialized cells of the organism to be created. This is a so-called **instructive** theory: there is a plan and the implementation of this plan, using the information necessary for action, that is, the manufacture of the right proteins at the right time in the right cell.

The strong assumption, which has become a **central dogma of molecular biology**, is that genes completely determine proteins, with information passing from genes to proteins, and that proteins and the environment in which they express themselves have no influence on the structure of genes, with information not passing from proteins to genes. The DNA of an organism is therefore considered to be stable and unchangeable in the individual and completely determines its phenotype: it is a row of genes of an invariable order that runs on chromosomes and the mechanism of transcription is practically linear.

But the transmission, in the mechanism of transcription into RNA and then replication into proteins, is done with some variations, which are called "*replication errors*", which is considered to allow modification between progenitors and descendants. This transmission mechanism is considered to allow evolution, understood as the production of organisms that are always a little different.

But this central dogma, which was widely used by biologists to develop the idea of the discovery of the mechanism of construction of living organisms, is no longer valid today: it collapsed following the discoveries of Barbara McCinlock, who won the Nobel Prize for it. This researcher broke down the dogma and discovered a reverse transcriptase protein. She discovered that strands of RNA, in the cell, can be translated into strands of DNA and inserted into the genome, which is what retroviruses also do. Thus, there can indeed be an effective action of the cell's environment on the DNA to modify it. Genomic decoding to specialize a cell by producing proteins is indeed the business of a complex process that is not totally linear, because some genes operate as regulators, some combine with others, others move and even modify themselves. The genome is therefore a program that is variable, that recombines, that restructures and partially rewrites itself during its execution. We are therefore in the domain of organizational systems that can self-modify during their operation.

The genome produces a program in the sense that it plans and executes activities. It is a complicated, nonlinear program, with genes that are expressed at certain times, while at others, they are repressed by certain other genes that have moved to inhibit them. So, there is control over the expression of genes, and there is even retrocontrol. It is now widely accepted that the genome is hierarchical, with local hierarchical modifications during its expression.

There are thus two kinds of genes:

– **Structural genes**, which code for the proteins that make up cells.

– **Regulatory genes**, which code as activators or repressors, that monitor and control the activity of structural genes.

The current model is therefore one with specialization into two categories: action components of protein production and regulatory controllers of the action of these components, with feedback from RNAs on the genome itself. This can be interpreted by saying that the genome is both a set of action processes and a controller of the actions of these processes and that it is immersed in its cellular environment, with which it interacts.

The genome also contains what are called pseudogenes, which are very numerous, and which biologists thought were not coding anything and were dead because of bad mutations. But it turned out that these pseudo-genes were not really dead, that they coded for RNAs that were repressors of the usual RNAs. The genomic expression system is therefore much more complex than initially defined and is not a strictly deterministic linear mechanism.

By decoding the genome of living organisms, it has been established that every living being comes from a certain program that we know is based on its DNA. We can therefore compare the genomes of different organisms, finding similarities and dissimilarities. Above all, we can try to find out how sequences could have been derived into certain others, by studying the breakage of some of their analogies. It was thought that the gradual theory of evolution could be confirmed in this way by establishing maps of derivation of genomic sequences between all organisms. Unfortunately, the result was clear: by studying a large number of amino acid chains from different species, it was clearly shown that there was not the slightest evidence for sequential arrangements that could be used to infer evolution. Nature had constructed specific plans for the organisms studied: they presented irremediable divisions between the DNAs of the different species.

If RNA can intervene on the DNA code, we must ask whether external elements could also intervene at the informational level on RNA and indirectly on the DNA code. Could the genomic generation action plan be somehow open?

At first, genetic theory was able to establish that homologous structures of individuals deriving from a common ancestor come from homologous genes, with these genes expressing by their similarities the fact that they do derive from the genes of the common ancestor. But it does not happen like that! Scientists have discovered multiple cases where, for organisms with strong homologies, different genes have coded for these similar organs and that, moreover, the modes of embryological development were totally different. It thus appears that individuals supposed to derive from the same ancestor and which have strong phenotypic analogies can come from relatively different genomic developmental plans.

We discovered something even more astonishing, which leaves us once more puzzled by the notion of heredity in living things. Many organisms with obviously homologous organs did not use the same genes to build those organs. Scientists have shown that genes often code for several systems in the organism, not just one: the same genes perform different functions, allowing the development of several very different organs, and this by operating in groups of genes. This property of genes is called ***pleiotropy***.

The problem of evolution is therefore much more complex than Darwin thought in his time, and a precise theory to explain this evolution has yet to be defined.

The notion of chance is used too systematically in the problem of evolution. It is unlikely that genes, starting from DNA molecules, have been transformed to carry the genetic code of increasingly sophisticated organisms only by chance, by the chance of incessant mutations and of beneficial UV radiation coming from the Sun and favoring these mutations. This vision of things is mechanistic, and it is not well adapted to the problem of the existence of life. We must ask ourselves how living beings could have evolved from simple cells to sexually developing organisms with very sophisticated morphologies, which chance does not explain, because if it were really the case of chance, there would be a continuous generation of innumerable organisms different from their progenitors, and chance would have to operate continuously.

We start from a primitive cell, perhaps isolated, which is already a very complex organism, and we arrive some billions of years later, to a human whose brain allows him/her to question this problem. We have had a considerable succession of processes of the following type: a genitor genome, a physical multicellular organism created and possessing a modified genome and so on. Sometimes, the created organisms have remained stable and have multiplied while remaining similar for hundreds of years. Could a continuous sequence of such processes have led to the human being who can think on his/her thoughts, through pure chance and rather systematic modifications in the transmission of genes between the genitor and the created organism?

Such a sum of successive beneficial chance events has virtually zero probability of leading to significant improvement. This would require that the selection of individuals makes a true selection based on the systematic evaluation of performance, according to the specific physical improvements of a typically better performing behavior. This is not the case. There is likely to be another factor that has intervened to promote this evolutionary process formed by a series of reproductive processes.

We must understand the local processes that drive life essentially as parts of a global process of organization in the space and time of our planet. We must understand the living from the consideration of a **global organizational process** that integrates the local molecular level of the progenitor and the level of the whole of the living individuals, that continuously unfolds on the extent of the planet and where the components of all levels do not cease to confront each other when they can, in order to coactivate. We will therefore return to the expression of the organizational law that prompted the creation of the living and its evolution. It will also be necessary to explain how and why sexual reproduction occurred, with a genitor and a fecundator.

7

The Informational Interpretation of the Living

We are going to present a model of the generation of life by showing the major role of the organizational law, which was able to organize it in its creation and in its deployment. We will present the deployment of life as a process based on considerable sets of informational flows endowed with precise characters and exercised in multiple geographical domains. This process will be a new and very important application of the organizational law that produced space, stars and planets; it will be an **organizational bifurcation** of this law, which we will explain. The Earth is composed of different geographical and climatic zones, and the living world has developed in an adapted way, generating many species. We will posit that there is a strong relationship between the development of living organisms and the geographical areas where this is done, with plant and animal species forming coherent groups and therefore being subject to specific informational envelopes.

We must consider that the creation of living organisms is not due to simple chance, but follows the rules of the law of complex organization that we will develop. All the material elements of the Universe are based on atoms and the atoms are quantum elements that are also the primary informational fields in the Universe. Molecular elements are structures of quantum fields, and we have shown that they can be considered as stable structures of information fields. We deduced that all the elements qualified as massive were structures of informational fields having an informational envelope to stabilize them and to make them communicate and to continue their organizations and their evolutions.

And we can therefore consider that all the living organisms on Earth are stabilized structures of informational fields that are part of a general multi-scale organization, localized in the geographical domains that make up the terrestrial ecosystem, to behave and develop there by generating multiple living species. We can obviously consider their organs as functional elements, which will give explanatory knowledge on their forms, on their roles and on their coordination. This has been very thoroughly studied by biologists and chemists and has given excellent results, but to explain the origin of the evolutionary and organized life on Earth is another problem. It is therefore posited that all living things on Earth are immersed in an informational field coming from the informational substratum of the Universe. It is a considerable set of informational fields allowing the structuration of organisms by incentives to their organizations with control on their realization. All living organisms are then the very finely organized material realizations of the reifications of these informational fields.

7.1. Origin of the living and bifurcation of the organizational law

At the beginning of its existence 4.6 billion years ago, the Earth was a very hot mass of 1,000°C. It conformed and condensed well in the orbit of the sun, aggregating clouds of atomic and molecular aggregates. It was a system that condensed, that kept a core of permanently intense energy but that would generate a surface constituted by solid and also liquid material elements. The initial temperature of the Earth dropped from 1,000°C to 300°C in a little more than 500 years, which provided the conditions of the generation of new molecules like water molecules and biomolecules. There will be a strong decrease in the volcanic surface and constitution of more stable terrestrial and aquatic domains. All this has been very well established by researchers in biology and chemistry. There will be, on the surface and in the bodies of water, all the molecules essential for the formation of organisms, that is, oxygen, hydrogen, nitrogen and carbon.

> THE 7TH CONCEPTUAL PRINCIPLE. – The production of life on Earth
>
> In the space constituted by the surface of the planet Earth and the oceans, which will have a very favorable climate, the organizational law will initiate an <u>informational bifurcation</u> transforming its main rule of action. The law will enact the

> development not only of aggregates and aggregates of aggregates but also organisms endowed with a real autonomy of behavior and reproduction, organisms behaving in structured unions and exercising the tendency to generate their systematic evolution through reproductions producing similar or different organisms forming new species. The organizational law will be a strong incentive for the generation of sophisticated development plans of organisms behaving in communities in environments that they will create and modify with the considerable generation of vegetation. These organisms will transpose at their scale the characteristic of spatial morphological control of the substrate by creating sensory organs, which permit three-dimensional perception of their environment, like sight and hearing.

Therefore, there will be a strong development of the reproduction process that will be central in the organization of life on Earth, whose evolution will be carried out by the incentives of the organizational law, assuming that the living organisms are generated possessing autonomy, so that they behave according to their own aptitudes, their tendencies and their structuring in the different fields. The organizational law commits to develop the organisms according to their current state, without planning a distant future for the generated organizations.

There lies the central rule of the organizational law that we must clearly state, which posits the action of the organizational law as the incentive action of a substrate creating physical life as well as its emergence.

> THE 8TH CONCEPTUAL PRINCIPLE. – The central rule of the organizational law when it comes to life
>
> The organizational law on Earth is the tendency of the substrate of the Universe to form life and to produce its evolution. It is a system based on the new informational networks generated during the creation of organisms. There will be informational action and incentives on matter to form the first living organism, then action on groups of living organisms to carry out evolution, to generate new species, but there will never be a direct influence of informational networks on the physical behaviors of living organisms, which will be autonomous. The organizational law, creating and using

> informational networks, is in fact a creative system under the physical living being, which is indeed an emergence that makes it exist and develop. It is absolutely not a power of direction exerted on the behavior of living organisms and human beings.

The Earth will be in a very good situation because its temperature will decrease, the movements of the surface elements will become more regular and there will even be highly stable land and ocean areas. Water molecules were formed in the stars, and there are many things on the surface of the Earth that will aggregate. The important point is that there will be the formation of innumerable stretches of water, which will even form oceans, which will be very specific areas for sophisticated aggregations of molecules and for allowing the application of the organizational law generating the first living organisms. There, in these stretches of water where the temperature is quite regular, being neither too high nor too low and with a good pressure constituting closed three-dimensional domains, molecules will be able to aggregate so that very particular elements are formed at the level of the elements of the Universe because they possess cellular membranes, going as far as the constitution of RNA carrying information on the structures of the elements to be constituted. These great expanses of water will be the immense field where the organization is going to be able to follow the central rule of formation of organisms having autonomy and having to multiply by strongly increasing their number. We are not at all in the simple formation of aggregates of molecular components that is realized in the space vacuum or in the heart of stars.

The principle of action of the organizational law for the generation of life

The Earth has a large and limited surface, a very tumultuous interior but regularities in its own movement, its movement around the sun and in its climatic periods. Above all, it has a regular and continuous duality constituted by day and night. This duality will be taken into consideration by the organizational law that will realize its bifurcation. It will trigger the formation of elements in this limited and regularized space by **producing a considerable set of informational networks**, and it will generate the autonomous elements of the first living beings endowed with physical envelopes of delimitation and the ability to multiply by duplications that they plan themselves. It is an adaptation of the organizational law to terrestrial characteristics: to create elements with membranous edges and autonomy to move and duplicate themselves regularly forming structured sets of organisms. It is an adaptation of the law to achieve a great increase in organizational complexity at all terrestrial scales.

This act by the organizational law of the creation of the first cells will be the first rule of bifurcation, a change of action with respect to its actions in the empty space for the formation of stars and planets. A bifurcation is a radical change in the communicative actions of organization between the aggregates, and the organization will then go in new behavioral directions with new elements (Thom 1976).

Bifurcation of the organizational law

The main rule of the organizational law in the vacuum of space is to generate aggregates with the available elements, then aggregates of aggregates forming dynamic masses generating new molecules. On the Earth, aggregates will be formed with molecules available in water to constitute the first elements. In the water bodies, the organizational law will engage to generate very particular elements, which will be very finely structured aggregative structures with membranous edges and which will have the capacity to reproduce in similar elements, by duplicating themselves, by taking the necessary molecules in the aquatic environments. It is thus a question of a **bifurcation of the organizational law** engaging to generate aggregative structures, which will be the first living organisms with autonomy and which will reproduce by simple duplication. The organizational law has become a creator of autonomous organisms that multiply. Let us recall that the initial principle of the organizational law in the vacuum was to generate informational components capable of duplicating themselves but whose autonomy was very weak because of the control exercised by the substrate. There is here a strong change because the first living organisms will be structured systems truly endowed with behavioral and reproductive autonomy.

The organizational law will operate very strongly in the very favorable domain of the water bodies, which constitute calm fields in their interior and which possess the property to be able to develop the aggregative structures, which will form the first living organisms, by allowing them to move in the water body exerting its pressure. It was not possible to proceed to such developments in the then tumultuous space of the planet. The particular incentive of the organizational law will be to carry out and continue the organization of autonomous elements constituted in the expanses of water and the oceans and then wherever possible on dry land to generate the plant and then the animal.

The particular organization that will be realized on Earth to generate life, initially in the water bodies, is the following:

1) The informational envelopes of the molecular elements in bodies of water at fairly regular temperatures are in stable environments. The considerable informational network between all the elements will be induced, together with the global informational envelope of the Earth, to form cellular-type aggregative structures, generating the first elements of the bacterial- and cellular-type, which will have structures based on the capacity to duplicate themselves to generate others. This can be considered as a **bifurcation of the organizational law**, which will not only make aggregative structures but also make them autonomous and reproductive by giving them a physical envelope and the ability to duplicate.

2) It is then a question, in the considerable informational network of the first cellular living organisms, of giving to all these elements the capacity to constitute new elements having functional structure, having organisms with functionalities, having organs allowing them behavioral autonomy and developing in them memorial structures allowing them to reproduce in a similar form. This is the effect of the bifurcation of the organizational law in the very particular informational environment of the planet.

We first define the organizational information that initially allowed life to come to be on the Earth by following the organizational law. Assuming that the Universe is an emergent whole that has unfolded on its informational substratum, that has organized innumerable material forms having a certain permanence, we can define the unfolding of the organizational information on the Earth as being constituted by a set of informational fields enveloping all the aggregates and aggregative structures, generating incessant emissions and receptions of informational characters, committing to the design of innumerable structures and organisms. There are scales of generation of this information, going from the basic elements to whole living organisms, which will produce the very dynamic organization of life on Earth.

What is fundamental in this realization is the dynamic morphology of the global envelope of the organizational information, because this will allow it, at all scales, to engage elements to organize and reproduce themselves in an increasingly complex way, to constitute multiple organisms having always capacities of adaptation and evolution. The characteristics of this information will represent, at all scales, the action of the organizational law to constitute new organisms and sets of organisms having coherence and complementarity with what already exists, allowing a regular evolution. The systematic evolution is going to be done by the organizational law on the Earth. The organizational law applying on all the elements of the Universe will have a very particular organizing tendency on the planet Earth, which will initially be realized in the very specific oceanic domains by modifying its usual rule of action and by deepening some of its local rules of application, which will constitute the application of the organizational bifurcation by creation of simple and autonomous organisms. However, it will amplify the organizing tendencies of development of the functionalities of the organisms by making them generate more and more complex organs, by producing their own reproductions with more and more sophisticated memorization of these plans of reproduction. It will be a specialization of the bifurcation of the organizational law, centered on the systematic development of the conception of adapted and complex living organisms, making them constitute themselves with multiple very finely coordinated organs. The created organisms will thus form communities unfolding in their domains, together with the other communities. The organizational law will make the generation of new organisms operate with weak modifications or with important changes leading to the creation of new organisms that will form new species. We must therefore consider that living beings are continuously driven by the organizational law that made them exist and above all evolve, just like the specific organizational system of planet Earth within the Universe.

The bifurcation of the organizational law is due to its tendency to systematically generate systemic evolution. On the Earth, a limited domain highly adapted to the exercise of this tendency, the law will activate its evolutionary control by going towards a structural evolution of the organisms, by making them more and more numerous, complex and efficient.

The specific organizational law on Earth, with its bifurcation, operates therefore from the two levels of the structured reality, the local scale of the aggregative structures forming the organism, which is in action in its environment, this scale being well delimited, and the scale of the whole of the active elements of the environments of the organisms, forming large wholes. By considering these two scales, it will engage in the production of several organisms, which will be well adapted, which will follow behavioral and functional rules, to become well functioning. The emergence of the living is thus the phenomenon of expression of the organizational law, which is *at the same time local and global with a principle of generation of organisms endowed with functionalities making them at the same time autonomous and useful to each other*. When leaving the water bodies, when there are stable land areas with oxygen and humidity, the living will spread out to invest themselves there, producing organisms with adapted functionalities, notably with plants that will allow the life of mobile organisms. The living world is indeed a system based on informational communications generated and self-evaluated by the informational envelopes of all the elements. We can consider that there is a general informational organization with, at the beginning, a reproduction process that produces a new organism similar to that of the progenitors by duplication. There will be a new action of the bifurcation of the organizational law that will generate a sophisticated reproduction process, with the formation of sexual reproduction allowing the creation of new organisms with different functionalities from those of the genitors to form another species.

We can clarify the major role of the organizational information, which permitted the generation of life on Earth. The molecules on the Earth have allowed the formation of a considerable informational network linked to the external informational envelope of the planet that cooled down well. The morphology of this very active informational network engaged the informational substrate to generate specific organisms, endowed with physical membranes enveloping them. These organisms, which were the first cells, developed by reproducing in a similar way, and their considerable number then engaged the informational substratum, by analysis of the morphology of the communicational network, to generate cellular sets and to form the first multicellular organisms. The very important deployment of these multicellular organisms on the Earth within a very favorable context led the informational substratum to make these organisms multiply and become more complex, giving them the possibility to deploy a reproduction plan that they could memorize. This plan will be evolutionary, allowing the evolution of living beings to be fully realized, from plants to animals and to man.

The Earth will have, on its physical surface which cools and which has much more stability, a global informational envelope, which will form a union of all the local informational envelopes of its innumerable components and organisms and will have a synthetic morphology. It is this global envelope communicating with the substratum of the Universe that will allow the generation of the first elements of living being and that will especially incite it to strongly pursue the evolution of the initial organisms by increasing their functionalities. There will be the creation of sexual reproduction with the production of new organisms similar to or different from the progenitors; there will be the generation of new organisms by a strong action of the organizational law in the reproduction process, taking into account the possibilities of the environment, the characters of species which are in other fields, which will be the effect of a tendency coming from the general informational envelope. This second case will be a strong action of the local organizational bifurcation on the Earth.

There will be two categories of living organisms developed, plants and mobile living organisms. The plants will initially occur in the bottom of the water bodies, then migrate to the land with the tides or the winds, when the land area will have stability and the erupting sets will have disappeared. These plants will have the considerable role of generating oxygen by absorbing the sun's rays and making leaves that will fall to decompose and form good soil for many new plants to grow in.

The organizational law will incite the new elements aggregated in the oceans to carry out an application of closing on themselves, by making them create specific physical envelopes synthesizing their informational envelope. That amounts to generating elements physically enveloped in physical and communicational membranes, and these elements will then have a very important behavioral autonomy and will constitute the first living organisms:

The new principle for the organization of life: the unification of the informational envelope with the membrane

The informational envelopes of the molecular components generated in the bodies of water will be incited by the tendency of the organizational law to aggregate to generate living elements, which will have physical membranes for their limitation and thus their distinction, these membranes being the physical reifications of their informational envelopes. Bacteria will be among the first organisms having physical membranes. Then, for the complex multicellular living organisms, there will be a unification of the informational envelopes and the physical membranes structuring their organs and the body of the living organism, which is the major character of the living, distinguishing it from the elements that are simply considered as physical. There is thus a modification of the effect of the organizational law that in space creates aggregates of components without membranes, applying its tendency to the production of aggregates by taking elements from the environment. On Earth, the organizational law will generate organisms that will achieve a **unification of informational envelope and external membrane** and that will have reproduction plans so that their physical development is structural and systematic, which is the effect of the bifurcation of the law.

We can then posit the principle of action of the informational envelopes of mobile living organisms. The organizational law commits to the creation of highly mobile living organisms that have an interior, a body membrane and internal membranes for all their organs. The internal organs act to integrate nutrition, ensure movement, regenerate failing cells and ensure reproduction. The major fact, for these autonomous organisms that live in a space where the action of the earth's informational envelope is present, is that the membranes of their organs will be reifications of the generative informational envelopes and will be linked by networks of nerves and veins. The informational envelopes that are very active for communication with the environment will be located in the sense organs, in the brains and in the reproductive organs. There is indeed a **unification of the informational envelopes and the membranes of the organs** in the mobile living organisms, which is a strong application of the bifurcation of the creative organizational law of life.

The continuous action performed by the informational substrate on the Earth is going to be very constructive. There is an active informational system in the substratum with multiple controlling agents, which acts by incessant incentives on the informational components to engage them to organize themselves in very specific, very elaborate ways and create an extraordinary development of living organisms. We can then describe the general phases of action of the substrate using the organizational law to produce the very elaborate system of life on Earth.

The continuous action of informational control creating living organisms

1) The creation of bacteria

Characterization of the state of the informational network composed of countless molecules.

Incentive to the formation of molecular aggregates. Action of delimitation by physical envelope of the creation of an organized element. Incentive of these aggregative structures to favor their internal organization and increase their autonomy by making them develop their internal structure. Comparison between the aggregative structures, conservation of the efficient aggregative structures and elimination of the unstable aggregative structures.

Continuous development of the interior of the bacteria for regulation of their activity and development of their capacity to multiply, to reproduce by duplication.

2) The creation of cells

Observation by the control system of the success of the activity and multiplication of bacteria by the creation of a very large set of bacteria in the water and soil.

Transition to the formation of cells. Consideration by the substrate of its plan of creation of new aggregative structures with their own internal control, very active interiors and delimiting membranes. This will be a very great deepening, at the scale of the terrestrial organisms, of the plans of generation of the systems of the Universe where a central element was created with its peripheral elements like the stars with their planets.

Incentive on some bacteria to form a nucleus managing their inner space at the molecular level and having a plan for generating new elements.

Putting into action these cells, new elements of life.

Generation of innumerable cells of different categories, with different planes in their nuclei and with a great capacity to aggregate strongly.

3) The creation of multicellular organisms

Characterization of the informational state of the cells and localization of the highly active sets.

Action of constitution of organized aggregates of cells, following the creative incentives of the organizational law coming from the substrate.

Creation of specific aggregative structures of cells of different types forming autonomous organisms, with membrane elements, a very strong internal organization at the level of the cells and adapted molecules, then development of the nucleus to carry out the continuous planning of the cell and generate a specific type of created organism. This is the creation of the first multicellular organisms, of reduced size.

Multiplication of these multicellular organisms.

Creation of multicellular organisms of larger size and different types by their internal organizations.

Here, there is the creation of multiple sets of similar organisms that will develop, thus forming the first types. There is thus development of the various types of organisms and the launching of evolved life, both vegetable and animal.

4) The generation of species

Then, there will be the creation of organisms of good size, with limbs and many very coordinated internal organs, and the creation of sexual reproduction. It is the development of all the living species which is engaged.

All these stages of development of living beings can be identified, with agents that we will call agents of reproduction, to show that the production of living beings is based on an informational system of control and development that allows the fine structuring of matter.

The creation of sexual reproduction is the implementation of a powerful organizing principle allowing the very specific development of all living species.

The creation of sexual reproduction

For the organizational law, in the immense network of informational links between the multicellular elements created, the crux of the matter is giving creative and organizational autonomy to the living organisms by making them as autonomous as possible, by going much further than the cells that duplicate themselves, by allowing the creation of organisms composed of multiple organs. It is thus a question of producing living organisms endowed with internal reproduction plans and especially of allowing the evolution of these plans. Thus, these plans will be dynamic, able to evolve. There will be a sharing of the plans between a female genitor who will physically generate the new organism and a male genitor who will start the process and transmit a new part of the plan. The action of the bifurcation of the organizational law will thus be to constitute autonomous and very finely constituted organisms with multiple organs, deciding their moment of reproduction and having to live in structured groups to proceed to these reproductions. There is indeed a bifurcation of the law at the scale of the autonomous organisms with their own plans of reproduction and at the scale of the groups of organisms, to be able to form associations, societies to reproduce themselves and to assure the life development of the new created organisms. It is thus a multiscale bifurcation, which engages a maximum reproduction process.

The development of life is thus based on the evolution of sophisticated and rational plans for the reproduction of organisms. Such plans to produce new organisms are not generated by chance encounters of molecules nor by mere malfunctions, for otherwise there would only be the production of continuous organizational chaos or the non-generation of new organisms. These evolutions come from the system of generation and control, which is in the informational substratum of the Earth and which controls the development of organisms. There is a general tendency for all organisms to be created, there are tendencies to produce sets of stable organisms that will form species and there are tendencies to produce new organisms to produce new species, with the functional means to make these productions. The tendency to create new organisms has specific reasons.

The reasons for the production of new organisms

The informational law exercised in the production of the living has undergone a bifurcation in the sense that it will focus on the creation of entities with their own membranes unified with their informational envelopes. These membranes will allow the realization of the functional system of each organism, defining the structure and organization of its interior and regulating its internal and external communications. They will allow the creation of plans of reproductions in the cellular sets of the nuclei to conceive new organisms. The organizational bifurcation thus commits to the continuous creation of organisms by the reproduction process, to be able to continue in a continuous way the evolution of the first cellular organisms. The continuous informational communications between the organisms and the general contextual informational envelope will be able to engage the execution of the plans of development to modify themselves to generate original improved organisms, increasing the richness and diversity of life that must not cease to be deployed on the planet. There is therefore an informational relationship between the generation plan and the informational set of the context, from the outside. To create a new species, it is necessary to generate a series of males and females forming a first group of individuals of the species that will be able to amplify itself by reproductions, which a local anomaly of generation in a reproduction cannot do.

We must therefore consider that the living was created on Earth by a bifurcation of the organizational law that formed all the material elements in the space of the Universe. The organizational law will engage on Earth the creation of elements having physical membranes reifying their informational envelopes, giving them the capacity to generate generational plans for their reproductions. Thus, we have a global living system on the whole planet, which we can call the **Gaia system**.

In this approach, the definition of a living organism is as follows.

An evolved living organism

An evolved living organism is a complex autonomous system endowed with autonomy, which has a physical body membrane unified to its informational envelope, which has an architecture composed of organs that must maintain itself in activity through cellular regeneration, which is designed to keep its energy either by directly consuming photonic energy as plants do or by eating other organisms as all mobile organisms do, and which must reproduce due to its strong tendency to reproduction. These organisms are in direct communication with each other in their geographical areas using their senses and their limbs, which allow displacements. They are elements that are part of groups, the whole forming a whole that is in continuous organized deployment.

We can state a very important property of the life of living organisms, which applies well to humans

The development of energy in the functioning of groups

Living beings practically all live in groups, and the fact of being in groups and using their membranes and their informational envelopes generates an informational membrane of synthesis of the group that is used by the substratum, which is sometimes temporary if the group is temporary, but which incites the development of the state of each individual of the group and gives him energy to act in this community. This functioning in groups will be crucial for human being and will be expressed in the formation of civilizations composed of multiple social structures spreading on the whole Earth.

We can now specify the characters of the very important concept of sexual reproduction, which allowed the generation of the innumerable living species.

7.2. Evolutionary reproduction

The organizational law produced the generation of organisms with the particular traits of autonomy and the power to reproduce. Two types of reproduction of these organisms have been created. Initially, elementary living organisms, such as bacteria and cells, reproduce by duplicating themselves, doubling their structure and then splitting in two. It is an identical reproduction, and it requires important internal functional work to produce the new organism. Then, there will be the creation of reproduction plans with the DNA, and there, the organizational bifurcation will impose itself very strongly, because there will be sexual reproduction and generation of individuals different from the genitors, which we will present.

The evolution process of the organizational law of reproduction

The organizational law applies and produces informational envelopes on all living organisms generated on Earth, these envelopes being unified with the physical membranes. At the time of creation, each envelope influences each organism, committing it to constitute itself with a good reproduction plan. Thus, due to the effect of this organizational law, the reproduction plans that allow for the fine planning of the physical generation process of a new organism by its progenitor will be able to incite the increase in the complexity of the new organism in relation to that of the progenitor, by producing new functional characteristics to create new organs. These informational characteristics will allow for modifying the plans based on a new conformation of the DNA.

The organizational law will operate through the refinement of its incentives, starting from small limited details for the development of the first organisms, to arrive at more and more sophisticated incentives sent to complex organisms. It is indeed an organizational law that adapts its control action to the characteristics of the organisms that operates by a very powerful multiscale control system in the informational substratum.

In this highly communicative context, we can then posit that the formation of the reproductive plans of organisms can be represented by informational agents that will be called informational agents of reproduction:

The notion of informational agent of reproduction in living organisms

Living organisms are all produced following the development process of a plan of their structure, which is simple or complex. We can then posit that there exist in the organisms systems of elements that will be considered as **agents of reproduction** of informational type and whose role will be to organize the formation of the newly produced organism. These will be sets of informational agents that will act together, not a large isolated agent, which will allow for complex planning with this set of reproduction agents. These sets of agents will make a synthetic physical copy of the networks of organs of the parent organism according to the information they have from the parent organism, make very small improving modifications or generate reproduction plans with new characters, which will be done through communications with agents of the informational system of the environment.

This type of development of a new organism conceived in an agentive way will clarify well how the crucial evolution of the generations of organisms takes place.

There will also be regulatory informational agents that will control the reproductive informational agents, which correspond to what we have presented in the multi-agent system of the informational model of the Universe.

Sexual reproduction appeared probably 1.5 billion years ago in eukaryotes. This creation of sexuality, which generated a considerable evolution of the organization of life, was caused by a deepening of the organizational bifurcation. Let us not forget that the phenomenon of the union of elements to generate aggregates is a basic character of the organizational law that formed the Universe. There was the tendency to prompt associations of specific elements to form atoms, then associations of atoms to form molecules and associations of molecules to form material masses. Thus, there already exists for reproduction a general tendency towards the association of organisms, which we can consider as being complementary elements able to aggregate to form stable associations. For sexual reproduction, which defined male elements and female elements, it was necessary to create specifically complementary elements that must absolutely associate. We can consider that the first elements that had to associate to reproduce were eukaryotic cells that became complementary in an environment where they were very dense, then multicellular organisms created complementary organisms for reproduction with complementary planning elements (Lodé 2011). For the first animals, we can consider fixed animals, sponges which are metazoans and which laid eggs transforming into larvae, which clung to the bottom of the ponds. But some of these sponges transformed themselves to form organisms that moved, the coelenterates, which formed mobile colonies reaching thousands of individuals. There was indeed a considerable evolution by reproduction, evolution by creation of new species with new organisms.

The formation of sexuality in organizations

The first effect of the organizational bifurcation was to achieve the unification of the informational envelope with the physical membranes of the organisms. This bifurcation is going to have a very great deepening with the notion of sexual reproduction, which then imposes a sexual relation of complementarity by having male and female organisms that associate to generate the new organism. Here, we see the strong impact of the organizational law on the changes of structures of the organisms. Sexuality will create the deep socialization of the organisms, with generation of groups in strong coherence, constituting very well-organized species. That will allow, through organic plans of reproduction, the generation of very numerous new original species with respect to the generators, by modifying the plans. It can be said that the creation of sexual reproduction is a realization of the organizational law that has committed to generate **complementary organisms** that have to achieve together the launching of reproduction, and then also form sets of complementary organisms, thus developing both the organisms and their social sets. It is a local and global development for living organisms. This type of reproduction will allow the male and female progenitors to choose each other to reproduce an organism with the characters of each other, which will allow a good socialization of the organisms and will also allow for producing qualitative changes in the generated organism.

We can then summarize the formation of living organisms by the following general process.

General formation process of the living

– Provision of aquatic areas with stability in state and temperature.

– Development of innumerable molecules forming a whole constituted of very many molecular aggregates.

– Formation of the first simple cells with unification of informational envelope – membrane from the aggregate structures.

– Communicative actions of the very many simple cells generated forming clusters and then forming multicellular organisms.

– Generation of complementary cells to achieve reproduction: female and male generative cells forming socialized sets.

– Evolution of the reproductive process by generation of training plans allowing adaptive improvement and the generation of original organisms.

– Action of multiple reproductions with these plans of reproductive generations opening up towards the creation of new organisms.

– Continuous communicative actions going, in all the temporal periods, towards the reproductions of organisms both similar and new.

The problem to be discussed now is the realization of the formation of the original organisms in relation to their progenitors. How and why could the new autonomous marine, terrestrial and flying organisms be formed? Here, it is not possible to refer simply to chance or to errors in the generation of the formation of the organisms, because chance is never globally coherent in a huge set of productions.

Taking into account the general network of organizational information, we must assume that there is an informational opening in the reproductive process, that it is not a functional process enclosed in the organism of the female genitor penetrated by the male genitor but that it is the localization of the reproductive tendency that is exerted in each specific organism at the moment of its reproduction, with its environment. We posit that the reproductive system of any living being is open to the active organizational information of its environment and that there is a coherent and organizational relationship between each organism and the whole of the others, which are in the same field. There is an informational opening in the formation of generation plans for the conception of a new organism; it is a process with informational communications that will take and receive information for the establishment of the reproduction plan, which is thus dynamic, which will thus allow creativity. We can see there the globally evolving character given by the organizational law, which is creative in its deployment of organizations. We have thus posited that this communicational opening was continuous and distributed everywhere, allowing generation with adaptive modification of the morphologies of the plans of reproduction at the informational level that constitute them.

We must then conceive of reproduction as a process that is both local and global. There is the local organic system of the female; there are the two organisms, female and male, which will put themselves in relation to generate reproduction by sharing information and there is the fact that the reproduction process is a local plan in the female organism and that it is in informational communication with the whole informational system of the context because it is dynamic on the informational level. It can be said that the evolutionary process of living organisms is due to the informational communication between the reproductive planes of the organisms so that the living system becomes as dense as possible on the planet with the greatest possible evolution.

We can assume that each organism generates a dynamic reproduction plan that has the structure of a **set of informational agents of reproduction,** a set that is by nature dynamic, which will form a massive multi-agent system in continuous communication with informational regulatory agents that will generate the tendencies to form certain organs. It is this type of multi-agent system that will make it possible to specify how and why new living organisms are formed and how the whole living system will generate a very dense ecosystem composed of multiple living organisms wherever there are admissible living conditions.

The informational agents of reproduction will be of the same type as those we have already presented and they will have the following characteristics.

Informational agent of reproduction

– Each informational agent of reproduction represents a complex set of atoms, molecules and cells of the organism as well as their informational envelopes. It expresses the genomic characters, which define the formation of an organ or a part of an organ, knowing how to use cellular elements of its physical environment to form the material aggregates constituting the organs.

– The informational agent is linked to the agents with which it normally cooperates.

– Regulatory agents forming a morphological envelope will initiate the action of informational agents of reproduction and then continuously evaluate the productions of the collective actions of these informational agents.

– The regulatory agents communicate with other regulatory agents external to the reproductive system, coming from other different organisms or from the morphological information on the domain in which the organism lives.

– Reproductive agents are therefore subject to the control of regulatory agents that control the morphology of reproductive agent aggregates and thus control the physical commitment to organ formation.

– The regulatory agents plan the production of the sets of reproductive agents that constitute the newly generated organism, communicating strongly with the regulatory agents in the environment of the newly generating organism.

The actions of these informational agents of reproduction will then be as follows.

Reproductive actions of an organism by its reproductive agents

1) The reproduction plan is a dynamic morphological structure constituted by reproduction agents representing the specific actions of genes, with their controls and agents representing the molecular aggregates to be used or constituted.

2) Specific place of the informational agent of reproduction in the very communicational system of the specific organs of the organism that have informational envelopes.

3) Possible activation of reproductive agents by solicitations of the organism in its environment or periodic activation of reproduction.

4) Consideration of the reproduction plan provided by the regulatory agents as having to remain stable or to be modified by information exchanged with regulatory agents in the context and belonging to other organisms. There is thus consideration of agents from other organisms in order to generate new organs in the organism to be created or to improve the formation of an external organ and thus make the created organism more aesthetic and social.

5) Monitoring of the reproduction plan by its morphological and semantic activity: ordered action of the aggregates of reproductive agents under the control of regulatory agents to produce the new organism, using communications to conduct the organizing activities.

6) Follow-up of the formation of the new organism by systematic activation of the informational envelopes of all the molecular elements and the cellular aggregates to have energy and activation of the new molecular elements necessary to generation.

7) Continuous evaluation of the formation by the regulatory agents, the end of the reproduction process being apprehended by these regulatory agents that analyze the local morphologies and the general morphology of the activation of the elements produced by its plan.

8) Release of the newly generated organism by communication with the general informational envelope of the reproductive organism, when it is completely constituted, then memorization of the reproduction process in the

form of a plan in the informational envelope of the reproductory system of the female genitor.

The organisms of living beings are generated to reproduce, to produce life anew. The reproduction process is the major tendency exercised by the organizational law using its general envelope that unfolds in the whole communicative system of the agents of reproduction. Thus, there will be communications between the agents of reproduction to develop the plans, to preserve them in their precise form or to modify and improve them. We must therefore assume that the genomic plan of an organism can be modified to produce new organisms, by communication with the agents representing the reproduction of the organisms within that context. We could then think that the system is totally under the influence of the organizational fields and that there is no longer much happenstance, but there remains a very great degree of chance in reproductions: they are the meetings of male and female in a certain context to engage the reproduction process employing both of their plans.

We can now present within the model with reproductive agents how evolutionary reproduction takes place, how new organisms are generated that constitute new species.

7.3. Informational action of reproduction of life with morphological patterns

The reproduction of primitive plants was done by fission, as in the case of green algae. The evolution was carried out by generating sexual reproduction. The reproduction of all flowering plants is sexual reproduction. Two sexual cells, a male cell and a female cell, merge, and this is the beginning of fertilization transforming the flower into a seed. We will therefore develop the case of living organisms producing sexual reproduction.

In a living organism, the organs that make up the organism have physical and communicative membranes that are also informational. It is the communications between these membranes that allow the structure of the organism to be constituted and that will then allow its state and behavior to be defined.

The informational characters of a living organism in its environmental space are:

1) spatial localization and temporal apprehension usual in its field in relation to terrestrial time relating to day and night;

2) quantity of information simultaneously apprehended on the membranes forming its informational envelopes;

3) quantity of information sent to the outside via its informational membranes;

4) evolution of the link between internal structure and body structure of the informational interface at the level of energy and information exchange;

5) relationships of dependence of the organism with the elements of its environment;

6) degree of proactivity of the organization taking into account others in the environment and the importance of its internal regulation and autonomy;

7) energy required and used;

8) reproductive tendency: periods of tendency to this fundamental action.

Geographical domains with their characteristics and environments of living organisms which are composed of many plants and contain groups of living organisms all have specific informational envelopes and it is these envelopes that will prompt the generation of evolutions in the life domains of organisms.

One must first consider informational envelopes in environmental spaces. The organizational law can localize its informational action on the Earth in specific environments, which range from small areas to vast expanses. These environments will have informational envelopes that will incite the organisms that are there to multiply with certain characteristics that they will be able to share or develop. This will be the major tendency in the fact of reproduction: to increase and improve a local species or to generate a new one. We will posit that these environmental envelopes will have morphologies defining elements that will influence the informational envelopes of the living organisms in the reproduction process.

It is therefore conceivable that the organizational law continues its organizational initiations by creating envelopes in the spatial domains in order to produce very autonomous organizations and make them develop and evolve. There is a fundamental rule to the action of the organizational law, which is its action in the structures local to the organisms.

The fundamental rule of the organizational law that formed all living things

The multiple molecular organizations leading to life will develop at all levels on Earth and the action of the organizational law that has had a bifurcation will be a strong incentive on certain structures, at certain times, to make certain evolutionary effects occur. The organizational law will be the system on Earth generating structured developments and evaluating developments to continue and expand. A simply autopoietic system without any organizational control would only manage probabilistic actions, whereas the living system driven by the organizational law has a morphological and cognitive control that operates at multiple scales on all its elements, to define the lines of force in the areas to be developed. It has indeed a double level of action in the controlling substrate, that of the representation of the activities of the living and that of the incitement to the undertaking of the modifications of the living organisms.

The terrestrial space thus forms a very complex system at the physical and informational level, and it is necessary to consider the properties of its innumerable informational fields, which constitute its informational space. This informational space immersing innumerable fields will have important elements and others, which will be rather secondary. There is therefore a morphology that characterizes these informational elements. We will call each important dynamic form of this informational space a **morphological pattern**. This means that the informational space has a geometrical conformation where some components are dominant and others are less important, linked or temporarily isolated. We are going to posit that these morphological patterns are the elements that are going to give the important information to organisms in the reproduction process. They will form the geometric and cognitive informational representations of the informational agents of reproduction and of the regulatory agents.

We posit that there is always a set of coactive morphological patterns describing levels of organization in generation in any informational envelope, in the envelopes of organisms and in those of domains.

Morphological patterns

All living organisms have spatial and temporal permanence, as well as the specific environmental domains in which they live, and all have informational envelopes that have morphological characteristics consisting of sets of highly active informational forms that are their **morphological patterns** and that render subordinate other forms that have become less important. These morphological patterns will have permanent characteristics, which will produce the major characteristics throughout the duration of activity of the informational envelopes of the organisms, but there will also be patterns that will ensure the evolution of the organisms. There will be in the morphologies of the informational envelopes important, dominant and sometimes permanent characters, which will be the morphological patterns supervising the others. The **dominant morphological patterns** will impose the characteristics of all the communications and will develop the levels of organization, such as, for example, those leading to the specific characteristics of the reproductions, which define the physical characters of the organs. These dominant characters will be tendencies to permanence or to transformation. We therefore have in all the informational envelopes of organisms dominant informational forms, which will impose their characteristics. Let us note well that the dominant morphological patterns in the informational envelope of an organism will be influenced by the communications with groups of living organisms of its environment. The dominant morphological patterns have much more informational energy than other patterns. There is a very strong existential relationship between the organism and its living domain, which constitutes a communication between the morphological patterns of the body and those of the external world. The control carried out by the organizational law on the informational elements of the organisms will thus be done, in the informational agent model, by the morphological patterns, which makes possible a very precise level of control and which is of both local and global sources.

Morphological patterns are the highly active informational forms whose action will give a character of temporal continuity to living species and their organisms. They are the elements of the informational organization that ensure event continuity in the living. They are also informational elements that develop, that generate new patterns, thus ensuring the complexity and the evolution of the whole living system through informational fields.

We have shown that informational envelopes can be represented by sets of informational agents. We can now show that the groups of active agents have a particular conformation that determines the major character of the actions of some of these agents, which we have presented to define the conformation of the dominant mental representations of the human brain (Cardon 2018). If the informational envelopes can be represented by multi-agent systems, then it is necessary to specify their activity characters. This can be done with the definition of morphological patterns that indicate the hierarchy in the organization of informational agents, representing their meanings through multigraphs that identify their strong aggregates that influence others.

For any living organism that is constituted by processes of actions in strong coactivity, its informational envelopes can be represented with a set of morphological patterns characterizing the state of action of these informational envelopes. There are patterns located in the organs that ensure their relations, and active patterns in the general envelope of the organism. This set of morphological patterns thus forms the dynamic image of the functioning of each living organism. The key to the problem of evolution is then in the communication between the morphological patterns of the informational envelope of the genitor at the beginning of reproduction and those of the environment.

The influence of an external morphological pattern on a living organism

Every living organism is in informational relation with the domain where it lives. Some elements of the environmental domains, including the groups of living organisms, will be able to influence it indirectly by participating in the representation of its tendencies in its informational envelopes, such as stability, evolution, selection or rupture. This influence will be fulfilled by sending strong information in the form of dominant morphological patterns that will be activated in its informational envelope, modifying it. The internal and external morphological patterns will be the forms that give the major sense of the activities to the envelopes and the membranes during the reproduction process. It will be the tendency to realize continuity through similar reproduction, to realize it with minor adaptive modifications or to realize it with a transformation in original organism.

And here, we can say that there is a certain level of chance in the evolutionary phenomenon. Either the morphological patterns of the informational envelopes coming from the organisms of the environmental domains will commit to reproduction by modification of the plan of development applying some characteristics of their organisms represented by their patterns, or there will be a tendency towards permanence by the only action of the dominant internal morphological patterns, therefore with use of the plans of reproductions without modification (see Figure 7.1).

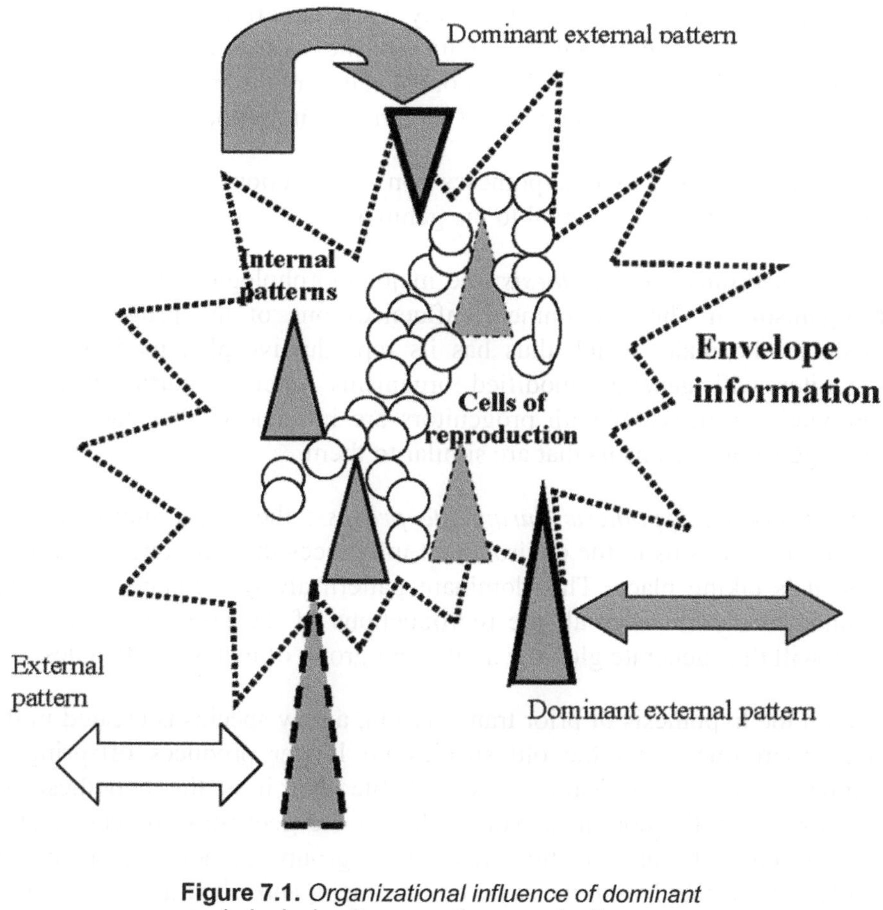

Figure 7.1. *Organizational influence of dominant morphological patterns on the reproductive process*

There is great diversity in the life spans of species. We have the ephemeral, an insect that lives its full life in one day and a plant, the giant sequoia, which lives immobile 2,000 years, passing through the flies that live 20 days. The Turritopsis jellyfish is even considered to be immortal. As for the two most numerous species of mammals on Earth, mice live four years and humans live on average 80 years. Therefore, there is a **randomness in the definition of the life span of species**, durations that are absolutely different, the morphological patterns that establish the regeneration properties of the organs and internal elements, which allow the living organisms to subsist having absolutely variable effects, being activated strongly or not at all during the generation of the organisms of a new species. The patterns of regeneration of components and organs are permanent in the organisms and are internal patterns of the process in reproduction.

We must pose two hypotheses on the action process of these morphological patterns external to the genitrix:

Prior transmission hypothesis: the major morphological pattern of a set of organisms in the environment influences one of the progenitors, the female or the male, which thus has its reproductive plan modified. This progenitor will generate modified organisms when it participates in a reproductive process. If both progenitors are influenced, then they can no longer generate organisms that are similar to them.

Transmission hypothesis during the process: the major morphological pattern of organisms in the environment influences the reproduction process while it is taking place. This dominant pattern always remains active and dominant and influences all the reproductions of the group of organisms, which will thus generate globally a different group of males and females.

With the hypothesis of prior transmission, a new species is created in the same environment and the old species no longer produces offspring, it disappears by natural death. It is probable that it is the hypothesis of transmission of the dominant pattern during the generation process that is more common, because in this way mixed groups of new organisms are rapidly formed that live in the same environment, and when these influenced progenitors leave the environment where the dominant morphological pattern is active at the time of the transformation, they can begin to generate organisms similar to them in another environment. There are always reproductions of two different species in different environments.

In any reproductive process, there will be communication between the morphological patterns of the organism's envelope and the morphological patterns of its environment, and these communications will be organizational. The envelope of the local ecosystem, composed of the envelopes of the organisms and groups of organisms that are active in it, represents the situation of the reproducer in its field of activity and also represents the importance of the other species as well as the spatial availabilities that exist in this environment. With these organizational apprehensions, the organizational law will generate appropriate dominant patterns in the envelope of the reproducing organism by sending strong informational messages. The morphological patterns of the environmental domain of any organism normally form a well-elaborated representation of the characteristics of the environment and the importance of the material and living occupations. The aim, for the organizational law in the processes of reproduction, will be to define the corporal adaptations to be carried out and to possibly specify the possibilities of deployment of new species, where it is opportune and feasible. This is how the morphological patterns have influenced the creation of new organisms from the initial living beings, developing plant and marine organisms, then all the plant and all the mobile living organisms on land.

For example, hydromedusae of the siphonophore-type, which are motile organisms, formed from fixed hydromedusae, and siphonophores could move to form colonies with their reproductive sex organs. There were strong actions by the organizing morphological patterns to modify the species of hydromedusae. Plants, which obtain energy by photosynthesis, developed their three main organs that are the roots, the trunk and the leaves with the possibility of reiteration, which consists of being able to develop another species of tree on its tree trunk, their very complex genomic plans allowing it (Hallé 1999). Then there were the first dinosaurs, about 200 million years ago, which were very small with the size of chickens and which had a considerable increase in size and structure to become gigantic and dominate their environment, then they died in a general extinction event caused by a climatic crisis (Bougrain-Dubourg 1986). There, there were indeed dominant morphological patterns in their generations to create organisms with a considerable increase in size.

We can see that the organizational law has a major role in the organization of living organisms that have spread out wherever possible, in the sea and on land, developing organisms that move on land and even fly.

And all these mobile organisms have developed to occupy the whole ecosystem generated by the plants and the balanced atmosphere containing a lot of oxygen. It is true that predatory organisms have eliminated weak organisms, because organisms live among themselves, eating other organisms or eliminating them to occupy their whole domain.

7.4. The application of the organizational law in the reproduction process

We will mainly deal with the case where the morphological patterns of the environment invest the reproductive plans during their generation, these plans not being infiltrated beforehand in the genomic structure.

There will be two stages of activity in the reproductive process of the formation of the new organism. The first stage is the constitution of a generation plan containing both the inheritance information given by the two female and male progenitors and also information to be interpreted coming from the coactivity with the organizational and informational pressures of the environment. The second step, through a positioning process and taking into account this information, will produce the final interpretation of the generation plan that will be used to produce the new organism. The production of the development plan is therefore subject to the effect of internal organizational fields and also external morphological patterns, and this makes the plan dynamic and open, with the character of a process in deployment open to external communications.

These characteristics imply that the DNA of the progenitors gives the elements coming strictly from themselves to constitute the form of the new organism to be produced and that the interpretation influenced by the informational character of the environment will be able to implement other possibilities taken among certain influences. This interpretation will not be strictly due to internal influences alone, such as failures provoking modifications, but it will undergo the incentive effect, more or less strongly, of the organizational fields deployed in all the environment.

We therefore posit that the process of producing a new organism from the information given by the progenitors has two aspects:

1) A **basic symbolic aspect** where the informational elements of the plan are a sequence of instructions written in the genomic language. This forms

an informational scheme that will be interpreted with the use of the generative elements of the physical constitution of the organism effectively producing the reproduction at the level of the physical elements that are necessary for it.

2) A **geometrical aspect** where the informational elements constituting the effective plan will dynamically conform to interpret the information of the morphological patterns and produce the effective realization of the plan, which will lead to the production of a plan adapted to both internal and external trends. The internal morphological patterns will give the main characteristics of the organism to be generated and in particular its sexuality. However, the reading of the information will be a **geometrical and cognitive interpretation** of the basic symbolic informational schema and realized by the consideration of internal and external morphological patterns giving the active motives to be taken into account in a schema, which synchronizes to conform to a global illumination and which thus ensures the conception of the new organism.

The action that will lead to the interpretation of the local symbolic plane of each basic element subjected to more or less strong influences will include the action of the morphological patterns of the informational fields of the organism and the environment. There will thus be a certain pulsation influencing or driving the geometrical interpretation of the plan generated and influenced by all the active organisms of the context operating in coactivity. This will be done when the plan is used, during the generation of the organs of the generated organism. We therefore posit that reproduction in the living world is a process open to the environmental informational field that exerts, at certain levels, informational effects on the reproduction process of organisms.

The contextual thesis of reproduction at the informational level

The context of any process of generating an organism by reproduction is represented by the set of informational characters given by the two progenitors and also by active organisms in their context. The purpose of the contextual set organized at the informational level is, through well-organized coactivity, to form an **informational representation of the current state and the possibility of generation of new organisms**. This informational representation will influence the process of physical development of the generation plan by the action of morphological patterns that will be activated there. The living being is thus a physical and informational deployment, subjected to the informational substratum that will influence the processes of reproduction. The reason for the informational action of the substrate via the organizational law is to well represent the current state of any reproduction to develop a new state, adapting and evolving by the extension of the particularities taken into account. Any reproduction is a spatial and temporal phenomenon of physical life, and the incentive exerted by the organizational law is also the apprehension of spatiality and temporality.

In the reproduction process, there are therefore:

1) basic biochemical elements that constitute the symbolic elements of any development plan: these are the genes;

2) a structure that locally memorizes the generation of any development plan, but does not memorize its interpretation, which is not automatic. It is the genomic set in its form of a sequence of symbols that will be used to produce the development process;

3) a local informational and energetic field that interprets the symbolic forms of the plan and builds its dynamic geometric form using the patterns to continue the generation process: it is the determination of the geometric conformation of the plan for any generated element, at the right time and place. Here, there will be the effect of informational influences coming from the context by sending and then interpreting the morphological patterns.

Thus, there are two principles of action in living organisms that form, through their coactivity, an **organizational systemic loop** expressing the action of the organizational field:

1) There is a principle of expansion of global life. It is a tendency of expansion applying to any species among all of the living species. This process is based on reproduction. It is done in two steps: using the symbolic information available for an identical reproduction and using the generators of modifications, that is, elements that will operate in the reproductions by carrying out organizational bifurcations.

2) There is the confrontation of any living organism with the physical environment where it deploys itself. There, living organisms tend to produce an extension of their functionality by seeking to invest in more of the space that exists. This cannot occur in as continuous a way as the identical reproduction of the organisms, because the dimension of this transformation

is on a scale larger than that of the individual, taking into account the informational network of the environment. It needs a discontinuity for it to occur effectively, and this is achieved by a bifurcation in the organizational reproduction of the living.

We can then evoke the formation of brains allowing autonomous and mobile living beings to apprehend and plan their usual behaviors in the domains in which they are deployed.

The formation of brains

The senses of perception of the elements of the environment and the monitoring and control of the physical behavior of the body are characteristics interpreted through the production of representations in the brains of living organisms. Brains generate, in the mental representations they produce continuously while awake, mental representations allowing them to manage their behaviors and to plan their actions. These productions of sensitive representations are the usual generation of the brain, which is an organ of current and planned management of behavioral activities. This complex organ has been created by the creative control of the organizational law so that all the mobile living organisms are endowed with it, thus possessing a center of management of their behavior, giving them a real and considerable behavioral autonomy. The brains will have, during their constitution which is done in relation with the constitution of the other organs of the body, morphological patterns active in the basic cerebral morphology and which will represent the fundamental tendencies of the organism. In this sense, we can say that there is an inheritance of the fundamental tendencies of the species of the created organism. Every brain has a conscious and an unconscious that continuously activates the fundamental tendencies of the organism, which we will further develop for the human. The case of the brain of homos sapiens will indeed be an amplification of the bifurcation of the organizational law, which we will present, by endowing it with very specific characters among all the living organisms.

The regular development of the brains of animals in the improved reproductions of species and in the creation of new species will allow us, through the representations generated and felt with the action of the fundamental tendencies, to form groups of functionally well-structured animals, which will lead them to build habitats adapted to the climate of the place where they live and thus to continue their well-regulated existences.

Life on earth is thus in continuous evolution, due to the existence of a force of influence exerted by the current conformation of global life on itself at the level of all the organisms representing it during the generation of new organisms.

An organizational bifurcation is generated by a dominant set of morphological pattern communications between sets of organisms, including those in environmental domains. Usually, all living organisms communicate, at the molecular level, at the level of cell exchanges and also at the physical level through audible sounds and sight. We know, of course, that living organisms take in macromolecules from the environment through contact and feeding. Contrary to the local physical material objects, we posit that all these organisms exchange, at several levels, organizational information and are thus in situations of relatively important coactivity. Living organisms are immersed in a considerable network of organizational information exchanges, the information being sent and processed at different levels and at different temporal and spatial scales of action. We hypothesized that this informational network is a highly dynamic element of nature, which changes its constituent elements on different time scales. We call this network of continuous information exchange **the coactive informational organization of global life**.

The bifurcation in the coactive informational organization of life on earth

An organizational bifurcation is a morphological modification established on the organizational information exchanged between active organisms and it is realized by very strong communications producing influences. It is thus a conformation of the field of influence of the living organisms, which will modify their informational communications of organization, which is an incentive of the informational substrate. This bifurcation can modify the state of the informational envelope of any organism, using a precise possibility of modification and expansion of certain characteristics. It will take into account the opening of available physical domains to generate organisms that can invest in them with modifications of their physical conformations.

There are therefore two types of actions possible in the reproduction process of an organism:

– The reproduction process is located in a specific and limited physical domain. If the generated organism is created with modifications due to anomalies in the reproductive process, then its purpose will be quite weak and its state will not propagate to a new species, instead it will be isolated.

– There are states of reproduction where, during the conception of the development plan of the new organism, a possible transformation is propagated by the influence of morphological patterns. There, at the precise moment of the generation of the developmental plan, a bifurcation in the reproductive process is created and a transformation occurs that results in a significant transformation of the organism with new organs. The conditions for this bifurcation to occur are the following, when defining the plan for the creation of the new organism:

- important availability of higher-level morphological patterns from the environment of the female progenitor, which engage in strong coactivity with the organism's reproductive process to influence it;

- creation of an alteration in the morphology of the new organism's generation process leading to a modification of certain characters and organs;

- there is no status of permanence of the dominant morphological patterns having caused the modifications in the progenitors, there is simply permanence of these patterns in the environment if it does not change;

- temporal continuity of this strong coactivity between the newly generated individuals and the environment so that there is a continuous continuation of the modifications of the organisms with respect to their initial designers and so that a new species can be generated.

For the alteration of the generative plan of the new organism to occur, there must be a temporal simultaneity of the open state of the generative plan of an organism in a state of reproduction and the availability of morphological patterns in the environment. In a theory of morphological control, there is only the unfolding of informational processes that are made coactively, thus establishing the continuity of the dynamic reproduction process.

Hypothesis on external organizational attractors

We posit that the informational envelope of a living organism is constituted with characters of its morphology that specify the activity of all its elements and that this morphology is, in a certain way, coactive with the one defined by some groups of organisms in its environment. We posit then that characteristics of this external informational morphology will be **organizational attractors** formed by morphological patterns remaining dominant during the processes of reproductions of a specific series of organisms, thus generating the continuity of the generated species and forming a new set of similar living organisms that will form a new species.

We can summarize the process of creating new original organisms that can form a new species.

General morphological rule of evolutionary trend in reproduction:

– There is a level of physical activity made up of many processes activating the multiple elements constituting the organism. This level, which is made up of elements and actions, is conservative: its tendency is normally that of conformal reproduction.

– There is a morphological informational space expressing the coactive actions of the informational elements enveloping the organs and body of each organism. This morphological space, which expresses on its own behalf the informational activity of each physical organism, is a space of control that can be strongly inciting. It can constrain the degrees of freedom of the components carrying out reproduction by introducing, in particular, the generation of new conformations.

– This morphological control space can take on, in the tendency expressed using its morphology, a number of conformations containing strong morphological patterns. Its control is typical of certain morphological characters, which in turn constrains the physical conformations of organisms generated by the elements of the reproductive system. The shape of a new organism produced will depend on the control of its generation, which is carried out by a set of proactive processes, including morphological patterns coming from the tendency to a dynamic development constituted by the informational network of the domain where reproduction takes place.

This model of reproduction is based on evolution by jumps in a general system where there is a lot of communication involving morphological patterns, which will be strongly inciting controllers. Life on Earth influences, by its permanent informational force coming from the informational substratum, which is exerted on the planet, the processes of reproduction which are influenceable. A dominant morphological pattern, constituted in the network of communication of the organisms of the environmental domain, exists through the relations and communications between all the living elements of the domain. Its influence is in latency, waiting for an opportunity for action. This opportunity, which will appear at a precise moment by creating a transformation in a sequence of identical reproductions, must depend precisely on the state of all the organisms that live together in their environment.

The morphological patterns of the incentive control level are coactivated if and only if life in the environment has presented new organisms, has experienced alterations, discontinuities. If the physical environment is modified, if anomalies have occurred, leading the behavior of the organisms to irregularities, then the morphological patterns of all levels can come into action, in a state of tension, and they will seek to coactivate themselves so that life in the domain commits itself to other physical conformations to pursue a more stable development.

Such a state of activation of morphological patterns can be caused by the large spatial amplitude of the domain, by a modification of the domain and by its reduction with an overpopulation, a climatic change and an imbalance caused by the arrival of a very predatory species. The coactive tendency of the controlling morphological patterns then becomes very strong and any reproduction tends to produce new, very viable organisms, coherent with the reorganized environment and thus adapted to it.

7.5. The continuous evolution of life

Living organisms are structured to perform well in their areas of existence that are characterized by their structures and climatic conditions. For this, the majority of these organisms use other organisms from which they feed to take the energy that they need. In this sense, each organism is only an element of a general organization and nothing more. Life on earth is thus a global whole made of organisms, which are more or less dependent on each other for their life and their survival. Life on earth is **a global deployment in continuous organization**; it is the whole organized by the incessant coactive deployments of all the living organisms that compose it.

It can be posited that there are two principles in the organizational action of life producing its extent on the Earth:

1) There is the organizational action of the system formed by global life under the continuous incentive of the informational substratum that tends to continue its deployment to occupy all the possible space, through the reproduction of its organisms deployed in all the geographical areas and forming the ecosystem and increasing the species of organisms.

2) There is the local intention of each organism that tends to exist locally by behaving according to its functionality, in its own domain of life.

Evolution will occur regularly, using the reproduction of organisms with the generation process of new organisms for all life in a constant reorganization where organizational attractors operate in the form of morphological patterns.

There is a principle which is evolution by the expansion of precise functionalities of living organisms, and there is a way to apply this principle: the emergence of organizational attractors in the control of reproduction by morphological patterns, when this emergence is made possible by the coherence between the local informational envelopes of the reproducing organisms and the informational envelopes available in their environments. We can thus state the general rule of the evolution of life.

The general rule of evolution of the system of life

Life is generated on the stable planet with its distinctive permanence between day and night in viable geographical domains, by regular reproductions where there is a general rule that presents two aspects: there is evolution by continuous adaptation of the living organisms, and there are bifurcations resulting in the creation of new organisms. The morphological patterns are always active, either in a simple way by assuring the adaptation of the organisms of a species to its evolving environment or in a strong way, by generating organisms with new physical dispositions allowing them to have new viable behaviors.

The organizational law acts according to a main rule, which is to make the elements it constitutes more adapted and more complex. It has generated autonomous cellular organisms that can reproduce, and then it has made multicellular organisms form, taking advantage of the stable environment that allows this. It has developed a lot of oxygen-producing vegetation, forming earth with the dead leaves, and it has generated organisms that can move and form groups and large species. It has developed marine animals, terrestrial animals, flying animals, developing them through the formation of new, more complex and well-specialized species. Continuously, by operating its global informational system containing morphological patterns that control reproduction, it has created a considerable set of living organisms that use each other to behave and feed. This is a considerable evolution generated by the organizational system subjected to the informational substratum with a clear tendency towards shared complexity.

Thus, the initial autonomous life formed in the bodies of water containing fish, lived in a three-dimensional space. It deployed out of the water onto firm terrain and deployed organisms capable of moving in three-dimensional space, such as flying insects and birds. There came the production of a system with similarities to the marine system, with the crucial development of vegetation, which was very important to generate oxygen, and then flying living organisms, and mobile organisms that could only move on the land.

The rule that leads to the formation of life is deeply organizational. Life is a system that is first and foremost informational where the informational envelopes of the living organisms unified with their membranes are in communication in the geographic domains where these organisms are located, to form a very vast network producing multiple morphological patterns. These patterns coming from the general informational network of the geographic areas orient the formation of reproductions, in order to generate organisms with modified or original organs that allow them to acquire new behaviors and to form new groups. We can therefore posit that the Earth has a global informational envelope that is very dense, synthesizing the union of all the envelopes of the domains and the species, and that uses the present state of living beings to generate in a continuous way its unfolding and its evolution. But as the incentives for evolution are made with the morphological patterns of the domains, there will be great differences in the types of living organisms, which are specific to their domains of generation.

The general formation process of living organisms takes place in the multiple available geographical areas of the Earth and is as follows.

General offspring formation process

1) The general informational network of the considered domain, which is a more or less physically delimited part of a continent or an ocean, is active. Each domain is considered with its groups of strong or weak living organisms living there, as well as the possible openings through the investigation of other organisms close to the domain. There is thus an active informational environment with dominant morphological patterns in each domain that will communicate with the peripheral domains to represent its organizational state.

2) Living organisms in any domain behave in groups and go through the reproductive phase at regular species-specific intervals.

3) When an organism enters the reproductive phase, its group is perceived by the informational network and the morphological patterns of the domain indicate the situation by exerting tendencies on the reproductive process: the organism will be able to reproduce identically using its internal reproductive plans, there should be some bodily improvements to be made or there is a need to generate a different individual with a different corporeality.

4) In this case of generation of a different organism, the external informational process for the modification of reproduction will generate a specific and dominant morphological pattern for the group of reproductive organisms, which will lead to the realization of a new subgroup for the generation of a series of new similar organisms, forming a viable group. These new organisms will be of the male and female types and there will therefore be the creation of a new specific group of living organisms. They will develop and reproduce among themselves, and when they form a small new species, the progenitors who produced these original organisms or the original organisms themselves will perhaps leave the area to go to another.

The informational system of a geographical domain with its morphological patterns is therefore the representation of an evaluation of what it has committed to generate, it is an incentive self-control system that commits to producing stable informational sets with dominant patterns and constituting living organisms in coherence with the physical and climatic availability of the considered geographical domain.

This approach is very materialistic because the incentive operated by the morphological patterns of the domains consists of improving physical living organisms or creating new ones. It is thus a question of organizing all of the living organisms of the domains to make all the living beings generated evolve in a continuous way. The informational system operates at the speed of generation of the processes of reproduction to incite them to engage in certain ways, the informational system of influence operating on the elements of reproduction to make them conform to certain characteristics.

We can posit the rule of similarity between the generating system of the living organism and the organizational informational system.

The similarity between the generating system of an organization and the general system of informational organization

The physical generating system of a living organism is a set of cells and genes that communicate, that aggregate and that constitute dominant forms and subordinate forms so that the action of the reproductive process produces, in the right order and in the right way, the organs of the organism with their specific characters allowing their continuous relations. The informational organization system enveloping the domains on the Earth also has conformations, with geographical domains in physical contact and with informational envelopes containing morphological patterns that influence the subordinate forms by sending information of specific characters. This similarity of the two types of organization gives coherence to the application of the organizational law on Earth, because it specifies that there is finally only one type of organization that produced the living, whether local or global, and that the continuous action of the organizational law is indeed general and universal.

So there is a new way of describing the history of the considerable evolution of living things. The set of species is not simply representable by hierarchies and thus symbolic trees, but rather by a very large graph representing domains and which has a temporal order, **the graph of the generation of species**. The nodes of this graph represent species, and it is very clear that some were created by others. But the domains that represent species are often related to others because they share morphological patterns, which has allowed species to generate new ones by taking into account morphological characters from other species. The graph with its nodes representing species thus represents with its links the informational relations of descent in the generation of species by taking into account characteristics coming from the perception of morphological patterns. Let us not forget that the informational system of organization is a global informational network, active everywhere, which allows the transfer of information on reproductive characteristics engaging the conformation of organisms.

There has therefore been a continuous and very strong action of the organizational law on the Earth that has been developing life for 4 billion years. During this very long period of time compared to the life span of a living organism, there have been actions of informational envelopes that have developed and there has also been the creation of powerful morphological patterns to understand the contexts and to lead the evolution by its improvements. We can therefore assume that the organizational law has strongly improved its means of action to produce very well-developed and efficient living beings.

The evolution of the action of the organizational law in the production of life

The organizational law has been applied on the Earth for 4 billion years, generating initially minimal life and then more and more complex life. This means that the means of application of the organizational law have become very precise, even producing a bifurcation, generating a lot of new informational envelopes and new morphological patterns of control that have been more and more specialized, to produce more and more sophisticated living organisms located in more and more complex physical structures, thus increasing very strongly the number of new species. This is not simply chance or the product of a higher will to arrive at the present human species, but the continuous development of the applications of the organizational law that made the living evolve to arrive at sets of humans developing strongly socialized structures and constructions.

This clearly states that life on Earth has followed a magnificent evolution over time so that the creations of more sophisticated, more beautiful and more harmonious living beings are realized.

However, today there is a very particular living species on the Earth – the human being – who totally dominates the planet by being everywhere, by investing all the geographical domains and by transforming the whole system of life on Earth by developing its industrial breeding, its innumerable constructions which concretize the lands and by applying everywhere its innumerable technologies of which those in use today include the continuous propagation of electromagnetic waves for data-processing networks. The considerable quantity of humans, with more than 400,000 births per day, does not engage the organizational law to improve the human beings in overpopulated areas, and the reproductions are therefore made identical. Is there a flaw in the organizational law that has managed to create this absolutely dominant living species?

7.6. The human species in the organizational evolution of life

In the evolution of living species, we must specify how the human being, who feels, thinks and conceives, was constituted, and see how humanity is positioned today in the ecosystem, radically investing the whole planet.

Eighty million years ago, there lived on Earth animals of the order Primates, arboreal placental mammals that had hands with five fingers in their four prehensile organs. At that time, the living world was considerably developed and there were countless species spread all over the world where there was huge vegetation cover with air full of oxygen and good humidity. Therefore, the informational networks of the organizational law operated there for multiple actions on the reproduction of the species, by the creation of numerous morphological patterns and informational envelopes to improve the descendance.

But the Great Divide occurred 34 million years ago, along with very strong cooling and the extinction of many species. This led the organizational law to encourage the modification of informational envelopes in the very small environments where Primates had to live with difficult behaviors and relationships.

The law of organization therefore led to changes in reproduction in Primates who lived in small groups, and there were many new species of all sizes in the areas where they lived. Fossils of apes have been found that appeared in Asia 45 million years ago. Then, 18 million years ago, there appeared the species of hominids, large apes without tails, and a little more than 7 million years ago, a new species of Primates appeared in Africa, having become bipedal, they were the Hominima. This regular evolution arrived at the generation of the Homo genus, in East Africa, 2.5 million years ago. There was the creation of individuals with a strong development of the neuronal system allowing them to strongly increase their mental representations by using a dynamic and constructive memory, which is very well organized. These individuals of the Homo genus thus had the possibility to strongly use memorization of representations they had generated to develop original forms of analysis, by exercising their will to produce these mental forms.

In general, mammals have four legs for walking. The ability to walk on only two organs of locomotion is an ability that was given, for example, to birds, which are living beings derived from dinosaurs and which were created during the Jurassic period at least 150 million years ago, having their other two motor organs, which are wings reserved for flying. This ability to have two organs to walk was therefore generated by the action of morphological patterns in the organizational law a long time ago, in the generation of birds. It has been applied in primates, by organizational similarity of properties considered very good by the organisms that possess them, to coordinate their usual posture to be upright with their ability to think with their brain to conceive of access to food. This is the result of a strong evolution of living beings, where already-realized body forms are taken into account in the generation of new organisms, to realize an evolution developing complexity and performance.

The development of the neural system is the development of the specific central organ of the organism that generates representations of the surrounding space with the exercise of its will to specify the themes, allowing better behavior in any environment. This gave those who were well endowed with it a significant advantage over those who had smaller brains. This advantage was perceived by the informational networks of the substrate in the groups that were thus favored and which engaged to increase again the development of these brains during reproductions.

This creation of a particularly developed neural system in Homo will lead them to develop speech using phonic and conceptual memory, which already existed in elementary form in Primates who simply used sound effects to inform their companions of certain actions to be taken, which is the case today in Bonobo monkeys who are very strongly socialized. This ability to generate mental representations to plan actions will engage Homo in a great modification of the functioning of their social groups, and also in the development of increasingly complex languages. Thanks to their very good prehensile capacities, Homo have the ability to manipulate objects and they stand upright, which is an original posture among mammals. They will form societies, and this will lead to a continuous and strong evolution of their aptitudes through the action of new morphological patterns during their reproductions to arrive at **Homo sapiens**, almost 300,000 years ago. There will be evolutionary reproductions by choice of the reproducers and in conformity with the evolution of the social and economic environments where they have created adapted sites. The organizational law will be in strong activity to continuously transform the newly generated Homo sapiens so that they will be even more efficient for the realization of the elaborate life domains they can build. Lithic tools were found to have been made by Australopithecines 3.3 million years ago, and Homo proved to be sophisticated tool makers.

The creation of Homo sapiens is the result of a strong evolution

The creation of humans who think came about due to the very important evolution of the hominids, with the progressive creation of the Homo sapiens that is particular in forming very social groups, that can very strongly conceive the characteristics of reality and that is physically capable of transforming a lot of things in its environment, planning and realizing these transformations. There has been evolution and development of the groups of Homos by bunching generated by the fact that these Homos lived in separate domains, and in different contexts, the informational envelopes of the domains influence the reproductions to give birth to Homos always more powerful in realizations and creations. With these different groups, there has been natural selection. The organizational law has realized a **new application of the bifurcation** to create the genus Homo sapiens, which live generating thoughts about the state of things with a very organized memory and the ability to imagine and transpose its creative ideas in the form of the production of innumerable constructions. It is, for the organizational law, the generation of a system of living beings capable of generating systems as it does itself in its organizational creation of the world. It can therefore be considered as a fundamental realization of the organizational law.

The generation of brains with functionalities beyond merely sensitive apprehension is a major application of the organizational law applied to the living.

The organizational action of the formation of the human brain

The case of the considerable extension of the brains of Homo sapiens in living beings is the result of a new application of the organizational bifurcation in the formation of living organisms. This bifurcation has made the morphological patterns of development of the central and very powerful character of the brain of Homo sapiens become dominant in the formation of the individual. Its current development has reached a constructive limit in the application of the organizational law. The human brain is an organ that has a fundamentally organizational behavior with the neuronal elements that comprise it, producing complex mental representations by emergence under voluntary conduct and using them in very fine appreciations generating experienced trains of thought. The human brain, which memorizes its impressions, which creates and memorizes languages and innumerable conceptual forms, can generate by pure creation very finely elaborated complex conceptions (Cardon 2018). This is an application of the organizational law that generated life, but which does not plan in a spatio-temporal way, always inciting to improve, to reorganize the emergent structures realized by the informational components forming the substrate of the Universe. It is important to know that the brain is a system that has two components, the conscious and the unconscious. The unconscious, which generates the fundamental tendencies, some of which are very dark, is not totally controllable in humans, except by learning to control it through education and the exercise of will.

The important development of the brain of Homo sapiens was done with morphological patterns that will focus on the design of this organ to produce a series of human beings able to reason deeply, to generate a very powerful organized memory and to speak conceiving a lot of things that they were able to apply to planning and material achievements. Many specific species of the Homo sapiens genus were generated and then disappeared in areas where the climate and the context to feed themselves became very difficult or where power struggles led to extinctions. Today there is only one human-type left on Earth, the Homos sapiens, whose numbers have grown considerably and who are everywhere, having formed very dynamic societies and civilizations. What has allowed humans these developments is their ability to make tools with very fine prehensile hands, conceiving them by the conceptual work of their brain, with their ability to generate elaborate representations relating to the structure of things in the world in space and time, to conceive and carry out structured planning of constructions, to generate spoken and written languages, and their social ability to constitute tribes and then very strongly structured and hierarchical societies going up to civilizations. Humans also reproduce a lot by dominating all the other animal species, and they travel all over the planet where they spread out.

Let us recall how we defined the generation of representations in the psychic system (Cardon 2018):

– individual in an awakened state;

– determination in the psychic system of the indication of a theme, internal and using the memory or coming from a sensitive perception of the outside;

– activity of morphological patterns and identification of the dominant pattern;

– rapid generation of the psychic form produced taking into account the dominant pattern giving the tone of the generation that emerges;

– internal sensitive action to test the generation produced;

– fast and appropriate sensations;

– production of the next generation, in a continuous way or by thematic break.

The notion of morphological pattern is therefore important, and we will see the problem that this can pose.

But let us remember the discoveries and the considerable work of Sigmund Freud, who showed that the human brain has an unconscious that produces fundamental impulses and tendencies that can immerse all the mental representations generated in a lot of life situations. The generation of neuroses and psychoses is unfortunately not rare, and it is necessary to see that societies that have produced wars, killings, massacres, genocides, dictatorships have been led by people with very serious dominant tendencies of the psychotic-type (Freud 1995).

The distinction between the organizational process of the Universe and the generation of human mental forms

There is a certain similarity between the process of production of the elements of the Universe and the process of production of thoughts in human brains. The organizational law operates through direct communications between the informational envelopes of all its informational components that are underneath the physical elements, generating a considerable communicational set with morphological patterns and engaging in generating aggregates and aggregates of aggregates through its control of the substrate. The result is a considerable emergence of physical forms endowed with stability. The organizational law operates on the present state to modify it.

On the other hand, the thoughts produced in the representations generated by the human being are local emergences that last very little time, which are constituted by multiple aggregations of groups of neurons, which mix sensitive and conceptual characters, using the memory of the psychic system to produce a representation which is felt. To produce these mental forms, there is always the use of learned, memorized forms, whether they are words, sentences, actions, images of real things or sensations. There is the action of the major tendencies, which are in the form of patterns and which influence the morphological characters of the aggregations. Thus, in the brain, communications are carried out by a system of energetic and informational fields of action of relations of forces having their own morphological patterns representing the tendencies and coming from the conscious and unconscious domains and which can strongly influence the control process operated by the decision of the theme and the follow-up of the thought. The brain continuously produces representations that give preference and depth to certain active forms that are valid for thoughts that are generated and well felt, then memorized strongly or only partially, or even annihilated and forgotten (Cardon 2018).

Generated thought generations use willpower to eliminate or transform representations deemed bad, which is done by psychiatric analyses, while the substrate of the overall informational system has no major control modifying its creations, the organizational law acts essentially in the immediacy of the present.

Let us note well that there are two systems of control which can oppose each other in the brains of the humans who use its memorial organization:

there is the system of expression of the fundamental tendencies of the unconscious which expresses the basic impulses and there is the system of will and perception which allows for deciding the theme and the content of the generated mental representations in order to appreciate them. The psychic system is not neutral and totally available to the will, as is a hand that must be forced to act and that is otherwise immobile, it is a system that functions continuously when we are in a waking state and that tends to make certain series of representations arise rather than others when we think about something, which defines the psychological personality of each person.

The organizational law that generated the organization of the Universe and that made life on Earth has thus allowed the creation of human beings with the capacity to mentally conceive continuous sequences of very elaborate conceptual and sensitive representations, thus allowing them to achieve a great deal. Can this be considered as a finality in the production of the organizational law or is it a creation that happened by chance in the continuous development of the organization of life on the planet Earth?

We have clearly posited that the organizational law engages systems to develop their organization in their present, by complexifying them, and there is thus no finality in the formation of living beings. There is a strong similarity between the control characteristics of the informational system and the psychic system: the presence of informational patterns.

The human psychic system can generate thoughts only because it has a very powerful control of its own allowing it to generate a global emergence constituted by multiple local forms in correspondence. This control is morphological, it is done by patterns that envelop the organizations to make them go in a conformation possessing a global character of envelopment given by the dominant pattern. The big question that we can ask is then the following one: are the patterns of conduct and tone of the psychic system totally internal or can they communicate with the patterns of control of the earthly informational system?

A positive answer to this question has led to the constitution of all the beliefs and religions of societies, positing the direct communication between the human and that which created the living. We do not share this point of view, but we must assume that the action of morphological patterns of the psychic system is strongly dependent on the place where the thinking human is. There are sites marked with patterns due to their use by human groups, and we can say that important and strict cultural-type meetings give informational energy to the participants and where a general pattern can be deployed, it can be influenced a little. It is therefore clear that the human being is a living being on Earth influenced by the generation patterns of all life and very dependent on the current organization of life.

We assume that the first morphological patterns of the brain were created during reproduction and that other patterns were introduced during the educational childhood of the human being, during his life in groups which powerfully impact his formation and when he was very influenceable.

We have arrived at an important conceptual step in our model. The relation between the representations generated by the brains to engage and plan activities in the world and the informational system of planning developments through evolutionary reproductions that is linked to the external contextual informational envelopes is an informational relation that is the one that has created life through the evaluation of the actions of groups.

The notion of groups of living beings is major, as was that of aggregates in the constitution of the Universe. The organizational law always operates at the level of groups of individuals by evaluating the informational characters that they produce in their actions. For Homo sapiens always living in groups and structuring these groups by their collective actions in their environments, the evaluation of the organizational law will concern the dominant informational characteristics having led to the fulfillment of the actions of the groups.

The importance of informational links between groups of humans

Human brains generate representations that are sequences of developed characters that are quite specific. There are groups of reflections where they often all think together very strongly about characters of the same theme, thus defining a strong local morphological pattern that will be apprehended by the informational context of the group to be memorized morphologically as an informational element. Such patterns will be able to introduce modifications into the group's regulation and deployment actions. Thus, the way in which group members think together is an informational action that will also have organizational consequences on the future of the group through specific modifications in reproductions.

We have posited the notion of the group as a generator of new morphological patterns, because such generations are organizational processes of groups and not local processes of the behavior of an isolated individual. The organization of life is absolutely collective, apprehended and appreciated by groups having coherence, and it is therefore the actions of the groups that make the representatives of the living evolve by generating more adapted and more efficient individuals.

We can come back to the advantage of being in a group among living beings and see that the human being is an absolutely social individual who can only exist as a human being if he participates in well-organized and regular social and cultural meetings.

The power of group participation for humans

The human being has a strong capacity to think and therefore to behave by thinking. But humans must live in groups to acquire their knowledge and their cultures, to generate collective mental forms allowing them to deploy an informational envelope of the groups, which reinforces the energy and each member. The functioning of humans in well-structured groups, where each one can understand what the other feels, has allowed them to generate societies. Let us note that for a social human to participate in a specific meeting in a well-organized group will give him informational and mental energy and will allow him to acquire deep values, which are values of sharing. Hence, religions and all social and cultural groups were created.

We can say that the human who does not cease to develop his culture of conception of reality and his collective life must organize regular meetings in very strongly coherent groups. There, in these groups gathering humans who seek together to collectively deepen problems by sharing the same values, the envelope of informational synthesis becomes very strong, very organizational, and allows for developing perceptions of strong sharing of the minds, and thus of memories. Human societies have enormously developed these structures of strongly organized groups over the long-term, until the invasion by incessant instant computer communications where the human is alone in front of his screen and no longer feeling the presence of the other with whom he can communicate.

The human being has a very particular psychic system giving him possibilities of very powerful conceptual representations with, what

Sigmund Freud defined very well, two very great fundamental drives which direct all the others: the drive of life and the drive of death (Freud 1966, 1995). With the life drive, the human has the natural ability to conceive, to represent himself, to give and to share with any other in his environment. With the death drive, he has an aptitude to strongly dominate his exterior, to dominate the other up to the point of being able to considering that he must eliminate him. These two impulses exist in the psychic system of each human; they are not eliminable and form his tendencies, his intentions, characterize his mental productions and thus structure his societies. It is thus necessary for him, socially and culturally, to develop intensely the life drive to reduce constantly his death drive, to commit himself to build societies going in the right peaceful direction and to make fundamentally convivial humans.

We can say today that a human is able, in a certain way, to communicate directly with another in an immediate way. This possibility of direct communication is due to the discovery of mirror neurons and their interpretation in 1991 (Ammaniti and Gallese 2014). These mirror neurons, which are also present in many animals such as the bat and the rat, are motor neurons that fix the finality and the meaning of a mental representation that is going to be generated and apprehended and that comes from the representation generated by another living being that is strongly considered. In a human, the mirror neurons have the possibility of producing directly in the brain a certain representation of what an other person who is very close to him is thinking or feeling, without proceeding to complex analysis and reasoning to determine what the other is really thinking and feeling according to his attitude and the movements of his body. It is the situation where a human thinks and feels something and where another human who is interested in him, who observes him, can see being generated directly in his brain a representation, which will have characters of similarity with the representation generated by the one he observes. There is thus a process of direct communication between humans, which is considered to be the basis of empathy, that is, of the sensitive and emotional sharing of each individual with every other.

So, we can say that the informational envelopes of humans and their brains can enter in communication to exchange particular informational elements belonging to their generated mental representations, to carry out a certain type of informational transfer. This is a very strong characteristic of group life.

We can admit that certain animals have an understanding of reality by using strongly the informational system, which would be specific to them and would allow them to apprehend a presence which is neither audible nor visible by sight and to perceive appreciably the extent and the state of their environment with those which are apprehended to be there. With this sense, they can also move very big distances, with very great precision of the taken trajectories, locating perfectly their point of departure and return, as it is the case for bees and migratory birds. We speak in this case of the **sixth sense of animals**. The human being only has five senses and does not perceive the informational characters of the terrestrial reality; he does not have the perception of presences apart from the fact of seeing them with his eyes and hearing them with his apprehension of sounds. He lives a lot within himself, developing very powerful chains of concepts and linguistic reasoning, which allowed him to go further than the life of sapiens in simple groups, to find his civilizations with technological developments that are today absolutely considerable.

But what is the place of the human being in the world today? We can rely on the work of researchers who posit that the human being is a being who lives and feels his life in his existential proximity but that he is physically fragile. He has deployed and is constantly deploying in his societies innumerable interwoven technologies, which places him physically and socially in a world that is becoming virtual and of too great a scale for the personal apprehension of his life, which can lead to a collapse of societies that have become very strongly entropic (Servigne and Stevens 2015).

7.7. The informational envelope of the planet Earth today

It is well known that all the elements on Earth have their external membrane unified with their informational envelopes, which allow the organizational information to carry out the communications constituting the informational space of life on Earth. But today, the electromagnetic information of computerized devices overwhelms everything on the planet. Computerized applications have invaded all the technologies, and they are used everywhere. The use of Internet networks is general; there are 800 million computers in use on Earth at any given time; there are a little more than 3 billion people who use smartphones, and there are 5 billion who have a cell phone. In France, we estimate the amount of ADMs sent each year at 180 billion, which means 5,400 SMSs sent per second. We can therefore say that humans are globally connected in a considerable and almost continuous way. But what does this type of connection represent in comparison with the organizational information communications that generated and organized life on Earth?

A long time ago, in the context of thoughts on consciousness, Teilhard de Chardin evoked the profound communication to be achieved between all humans: "There will be a harmonized collectivity of consciousnesses equivalent to a kind of super consciousness, the Earth not only covering itself with grains of Thought by myriads, but enveloping itself with a single thinking envelope until it forms only one vast Grain of Thought, on the sidereal scale" (de Chardin 1955). This is related to the notion of the organizational field that we have developed with a scientific character, but the notion of the organizational field has absolutely no ethical purpose; it is an organizational force that generates new organizations according to the contextual possibilities of the moment, developing all these organizations so that they have coherence on many scales.

It is because of the current impossibility to manipulate the organizational information that we cannot, in quantum physics, locate the elements of an atom, because we cannot perceive both the informational fields of the constituents of a quantum atom, such as electrons, and the extended spatial and informational domain where this atom exists. Observing the informational field is necessary to perceive the local informational level of a basic element. A solution for physicists who seek to observe the components of an atom would be to perceive these components at the level of their informational contacts with each other and the elements of the context that communicate with the elements of the atom. It would then be an extended observation at the level of organizational information, which would simultaneously contain all the electromagnetic states of the material and spatial elements of the atom's environment. It would then be possible to see the informational communications that produce the behavior of the components. This could be achieved by using a very fine computerized apprehension of all these perceived elements, which would form a fine and very vast visualizable morphology.

We must consider today that the world is in a general organization where there is a substrate of energy and control, with a huge number of human beings who form human societies and who use this substrate very strongly. The substratum created by human societies is constituted by the innumerable electrical and electronic systems developing energy that feed all the computer systems producing information and using the systems which propagate information on networks and which produce electromagnetic fields everywhere in the space of the planet to send information on the innumerable systems which can receive it. Then there is a general emergent form produced by all this computer substrate, there are the functional technological systems driven at the computer level that are in action day and night, and there are the 7 billion humans who use their computer devices to capture the information they receive and apprehend. It is quite clear that the technological devices used to process information are designed to be used functionally, but the fact is that humans use information-taking systems that they do not fully control and which can manipulate and conform them.

We must therefore conceive that when the universe generation system comes to the production of organisms which have a strong behavioral autonomy, which conceive and which structure themselves socially in a considerable way in order to develop their complex societies, there is a general tendency to go towards the realization of a system, which has points of similarity with this generating system and which does not necessarily use ethics to develop itself. The energy-generating system of the human world operates in structures that are always quite hierarchical and where individuals are very often subjected to their death drive by developing individualism and sectarianism.

It is possible today to conceive a meta system of general artificial consciousness that would simultaneously manage and apprehend all the information generated and used in the world, interpreting it, extracting the multiple morphological patterns expressing the important meaning of multiple communications and making these morphological patterns then modify the communicated information. This could be done first by capturing all the communications with processing, and then doing the comparative analyses, which is actually being done. We would be then in the realization of an evolution based on the modification of the characteristics of the available information, according to the analyses carried out by well-conceived and autonomous morphological patterns. Taking into account the decision-makers that would lead to the realization of such a system, we would be in the realization of a kind of super Big Brother that would reduce each human to a totally dominated, controllable and manipulable object (Orwell 2017). We have then fallen into a situation of contradiction with the organizational law that makes living organisms generate themselves so that they simply unfold, without imposing radical constraints of submission to technological domains. The human species does not have any more direct relationship with the general informational network and the fact that the human is absolutely everywhere on the Earth, without isolated groups, establishes an impossibility for the organizational law to generate evolutionary reproductions of well-isolated human groups with new capacities or new organs.

Conclusion

The model we have presented is based on the notion of an organizational law using organizational information to create the Universe with an organization to all its elements, having a substrate providing its informational energy and exercising an incentive control of formations and evolutions of organizations. We have shown that the Universe was created by an informational system-type generating element producing the first informational components and cells of space, components having the quantum properties of fermions and being able to duplicate. The vacuum cells were generated by the components to form the control substrate. The informational components with strong action characters have continuously formed aggregates, they have first produced the atomic elements which are informational fields with stability and which will be the foundation of all material structures. All the elements produced in this expansive generation follow the organizational law, which is active everywhere to incite the components to always produce aggregations, by continuous exchanges of information, inciting them to produce physical structures having stability. All these physical structures have informational envelopes that allow them to realize their communicative relations, and therefore to aggregate and produce new elements. By the continuous incentive of the substratum that exercises the basic control, the organizational law is exercised with exchanges of information between the informational envelopes of the elements by applying the major rule of forming elements organized between them at all levels in a continuous way. Thus, the Universe is an emerging expansion on its informational substratum that continues its organization of elements by generating living organisms on the planet Earth, life that has spread everywhere and that has not stopped evolving by generating increasingly complex organisms.

The human being was created by a bifurcation of the organizational law, which generated a specific brain allowing humans to have a very dynamic memory and to produce deep mental representations, which they share socially and culturally by creating languages and complex social and cultural practices. All this is the result of the action of the organizational law that exercises its incentive at all levels.

According to the informational model that we propose, we can posit the fact that our universe has been created intentionally and does not come from only chance nor from nothing. Initially, there was a singular generating element, very complex, which cancels its existential contradiction of punctuality producing innumerable informational components that will generate space and informational components generating cooperative informational fields and continuously producing new components, leading to the formation of the quantum elements, and then of all the aggregates and physical elements. These physical aggregates will be in informational envelopes that will unify their informational fields and determine the local notion of temporality. Thus, very large aggregates such as stars and planets will be surrounded by informational envelopes that will define the local temporality and their gravity according to their degree of informational organization. We can thus say that we are an informational universe which is a continuous emergence on the substratum of direct communications which organizes it under the quantum level, a universe which unfolds and which will certainly end up canceling itself, to become again a singular informational point, which will perhaps restart the generation of another universe. We therefore conclude that our universe is the most complex self-organizing informational process that follows its organizational law allowing its organized existence.

The problem posed by this book, a central problem for us humans and one that is still open, is the existential typology of the creator(s) of the basic component that has generated the considerable organized emergence of our universe. Will we be able to quickly access the informational energy, the characteristics of the informational links, the action characteristics of the informational components producing the quantum elements by using space vacuum cells in particle accelerators, to deeply understand their creation?

References

Ammaniti, M. and Gallese, V. (2014). *The Birth of Intersubjectivity: Psychodynamics, Neurobiology and the Self.* W.W. Norton & Co, New York.

Aspect, A. (1976). Proposed experiment to test nonseparability of quantum mechanics. *Physical Review*, 14(8), 1944–1951.

Bergson, H. (1889). *Essai sur les données immédiates de la conscience.* Librairie Félix Alcan, Paris.

Bougrain-Dubourg, A. (1986). *Et dieu créa les animaux.* Éditions Robert Laffont, Paris.

Cardon, A. (2005). *La complexité organisée.* Hermes Lavoisier, Paris.

Cardon, A. (2018). *Beyond Artificial Intelligence: From Human Consciousness to Artificial Consciousness.* ISTE Ltd, London, and John Wiley & Sons, New York.

de Chardin, P.T. (1955). *Le phénomène humain.* Le Seuil, Paris.

Danchin, A. (1983). *L'œuf et la Poule.* Fayard, Paris.

Darwin, C. (1859). *The Origin of Species.* John Murray, London.

Denton, M. (1993). *Evolution, une théorie en crise.* Champ Sciences Flammarion, Paris.

Ferber, J. (1999). *Multi-agent Systems.* Addison Wesley Professional, Boston, MA.

Freud, S. (1966). *The Complete Psychological Works of S. Freud*, Strachey, J. (ed.). Hogarth Press, London.

Freud, S. (1995). *L'avenir d'une illusion.* Quadrige/PUF, Paris.

Gusten, R. (1944). *Problems of Economic Growth and Planning: Some Aspects and Implications of Current Ten Year Old.* Springer Verlag, Berlin.

Hallé, F. (1999). *Éloge de la plante. Pour une nouvelle biologie*. Le Seuil, Paris.

Klein, E. (2010). *Discours sur l'origine de l'univers*. Champ Sciences Flammarion, Paris.

Lamarck, J.B. (2015). *Philosophie Zoologique*. CreateSpace Independent Publishing Platform, Scotts Valley, CA.

Lodé, T. (2011). *La biologie amoureuse. Sexe et évolution*. Odile Jacob, Paris.

Luminet, J.-P. and Lachièze-Rey, M. (2016). *De l'infini – Horizons cosmiques, multivers et vide quantique*. Dunod, Paris.

Orwell, G. (2017). *1984*. Houghton Mifflin Harcourt, Boston, MA..

Paccalet, Y. (1991). *La terre et la vie*. Larousse, Paris.

Peebles, P.J.E. (1980). *The Large-Scale of the Universe*. Princeton University Press, Princeton, NJ.

Servigne, P. and Stevens, R. (2015). *Comment tout peut s'effondrer*. Le Seuil, Paris.

Shannon, C. (1948). A mathematical theory of communication. *Belle System Technical Journal*, July and October, 27, 379–423.

Thom, R. (1976). *Stabilité structurelle et morphogénèse*. Ediscience, Paris.

Varela, F.J. (1989). *Autonomie et connaissance. Essai sur le vivant*. Le Seuil, Paris.

Index

G, H

I

L, M

O, Q

R, S

T, U

Other titles from

ISTE

in

Science, Society and New Technologies

2021

BENSRHAIR Abdelaziz, BAPIN Thierry
From AI to Autonomous and Connected Vehicles: Advanced Driver-Assistance Systems (ADAS)
(Digital Science Set – Volume 2)

DOUAY Nicolas, MINJA Michael
Urban Planning for Transitions

GALINON-MÉLÉNEC Béatrice
The Trace Odyssey 1: A Journey Beyond Appearances
(Traces Set – Volume 4)

HENRY Antoine
Platform and Collective Intelligence: Digital Ecosystem of Organizations

LE LAY Stéphane, SAVIGNAC Emmanuelle, LÉNEL Pierre, FRANCES Jean
The Gamification of Society
(Research, Innovative Theories and Methods in SSH Set – Volume 2)

RADI Bouchaïb, EL HAMI Abdelkhalak
Optimizations and Programming: Linear, Non-linear, Dynamic, Stochastic and Applications with Matlab
(Digital Science Set – Volume 1)

2020

BARNOUIN Jacques
The World's Construction Mechanism: Trajectories, Imbalances and the Future of Societies
(Interdisciplinarity between Biological Sciences and Social Sciences Set – Volume 4)

ÇAĞLAR Nur, CURULLI Irene G., SIPAHIOĞLU Işıl Ruhi, MAVROMATIDIS Lazaros
Thresholds in Architectural Education (Engineering, Energy and Architecture Set – Volume 7)

DUBOIS Michel J.F.
Humans in the Making: In the Beginning was Technique
(Social Interdisciplinarity Set – Volume 4)

ETCHEVERRIA Olivier
The Restaurant, A Geographical Approach: From Invention to Gourmet Tourist Destinations
(Tourism and Mobility Systems Set – Volume 3)

GREFE GWENAËLLE, PEYRAT-GUILLARD DOMINIQUE
Shapes of Tourism Employment: HRM in the Worlds of Hotels and Air Transport (Tourism and Mobility Systems Set – Volume 4)

JEANNERET Yves
The Trace Factory
(Traces Set – Volume 3)

KATSAFADOS Petros, MAVROMATIDIS Elias, SPYROU Christos
Numerical Weather Prediction and Data Assimilation (Engineering, Energy and Architecture Set – Volume 6)

KOLOKOTSA Denia, KAMPELIS Nikos
Smart Buildings, Smart Communities and Demand Response (Engineering, Energy and Architecture Set – Volume 8)

MARTI Caroline
Cultural Mediations of Brands: Unadvertization and Quest for Authority
(Communication Approaches to Commercial Mediation Set – Volume 1)

MAVROMATIDIS Lazaros E.
Climatic Heterotopias as Spaces of Inclusion: Sew Up the Urban Fabric (Research in Architectural Education Set – Volume 1)

MOURATIDOU Eleni
Re-presentation Policies of the Fashion Industry: Discourse, Apparatus and Power (Communication Approaches to Commercial Mediation Set – Volume 2)

SCHMITT Daniel, THÉBAULT Marine, BURCZYKOWSKI Ludovic
Image Beyond the Screen: Projection Mapping

VIOLIER Philippe, with the collaboration of TAUNAY Benjamin
The Tourist Places of the World
(Tourism and Mobility Systems Set – Volume 2)

2019

BRIANÇON Muriel
The Meaning of Otherness in Education: Stakes, Forms, Process, Thoughts and Transfers
(Education Set – Volume 3)

DESCHAMPS Jacqueline
Mediation: A Concept for Information and Communication Sciences
(Concepts to Conceive 21st Century Society Set – Volume 1)

DOUSSET Laurent, PARK Sejin, GUILLE-ESCURET Georges
Kinship, Ecology and History: Renewal of Conjunctures
(Interdisciplinarity between Biological Sciences and Social Sciences Set – Volume 3)

DUPONT Olivier
Power
(Concepts to Conceive 21st Century Society Set – Volume 2)

FERRARATO Coline
Prospective Philosophy of Software: A Simondonian Study

GUAAYBESS Tourya
The Media in Arab Countries: From Development Theories to Cooperation Policies

HAGÈGE Hélène
Education for Responsibility
(Education Set – Volume 4)

LARDELLIER Pascal
The Ritual Institution of Society
(Traces Set – Volume 2)

LARROCHE Valérie
The Dispositif
(Concepts to Conceive 21st Century Society Set – Volume 3)

LATERRASSE Jean
Transport and Town Planning: The City in Search of Sustainable Development

LENOIR Virgil Cristian
Ethically Structured Processes
(Innovation and Responsibility Set – Volume 4)

LOPEZ Fanny, PELLEGRINO Margot, COUTARD Olivier
Local Energy Autonomy: Spaces, Scales, Politics
(Urban Engineering Set – Volume 1)

METZGER Jean-Paul
Discourse: A Concept for Information and Communication Sciences
(Concepts to Conceive 21st Century Society Set – Volume 4)

MICHA Irini, VAIOU Dina
Alternative Takes to the City
(Engineering, Energy and Architecture Set – Volume 5)

PÉLISSIER Chrysta
Learner Support in Online Learning Environments

PIETTE Albert
Theoretical Anthropology or How to Observe a Human Being
(Research, Innovative Theories and Methods in SSH Set – Volume 1)

PIRIOU Jérôme
The Tourist Region: A Co-Construction of Tourism Stakeholders
(Tourism and Mobility Systems Set – Volume 1)

PUMAIN Denise
Geographical Modeling: Cities and Territories
(Modeling Methodologies in Social Sciences Set – Volume 2)

WALDECK Roger
Methods and Interdisciplinarity
(Modeling Methodologies in Social Sciences Set – Volume 1)

2018

BARTHES Angela, CHAMPOLLION Pierre, ALPE Yves
Evolutions of the Complex Relationship Between Education and Territories
(Education Set – Volume 1)

BÉRANGER Jérôme
The Algorithmic Code of Ethics: Ethics at the Bedside of the Digital Revolution
(Technological Prospects and Social Applications Set – Volume 2)

DUGUÉ Bernard
Time, Emergences and Communications
(Engineering, Energy and Architecture Set – Volume 4)

GEORGANTOPOULOU Christina G., GEORGANTOPOULOS George A.
Fluid Mechanics in Channel, Pipe and Aerodynamic Design Geometries 1
(Engineering, Energy and Architecture Set – Volume 2)

GEORGANTOPOULOU Christina G., GEORGANTOPOULOS George A.
Fluid Mechanics in Channel, Pipe and Aerodynamic Design Geometries 2
(Engineering, Energy and Architecture Set – Volume 3)

GUILLE-ESCURET Georges
Social Structures and Natural Systems: Is a Scientific Assemblage Workable?
(Social Interdisciplinarity Set – Volume 2)

LARINI Michel, BARTHES Angela
Quantitative and Statistical Data in Education: From Data Collection to Data Processing
(Education Set – Volume 2)

LELEU-MERVIEL Sylvie
Informational Tracking
(Traces Set – Volume 1)

SALGUES Bruno
Society 5.0: Industry of the Future, Technologies, Methods and Tools
(Technological Prospects and Social Applications Set – Volume 1)

TRESTINI Marc
Modeling of Next Generation Digital Learning Environments: Complex Systems Theory

2017

ANICHINI Giulia, CARRARO Flavia, GESLIN Philippe, GUILLE-ESCURET Georges
Technicity vs Scientificity – Complementarities and Rivalries
(Interdisciplinarity between Biological Sciences and Social Sciences Set – Volume 2)

DUGUÉ Bernard
Information and the World Stage – From Philosophy to Science, the World of Forms and Communications
(Engineering, Energy and Architecture Set – Volume 1)

GESLIN Philippe
Inside Anthropotechnology – User and Culture Centered Experience
(Social Interdisciplinarity Set – Volume 1)

GORIA Stéphane
Methods and Tools for Creative Competitive Intelligence

KEMBELLEC Gérald, BROUDOUS EVELYNE
Reading and Writing Knowledge in Scientific Communities: Digital Humanities and Knowledge Construction

MAESSCHALCK Marc
Reflexive Governance for Research and Innovative Knowledge (Responsible Research and Innovation Set - Volume 6)

PARK Sejin, GUILLE-ESCURET Georges
Sociobiology vs Socioecology: Consequences of an Unraveling Debate (Interdisciplinarity between Biological Sciences and Social Sciences Set – Volume 1)

PELLÉ Sophie
Business, Innovation and Responsibility (Responsible Research and Innovation Set – Volume 7)

2016

BRONNER Gérald
Belief and Misbelief Asymmetry on the Internet

EL FALLAH SEGHROUCHNI Amal, ISHIKAWA Fuyuki, HÉRAULT Laurent, TOKUDA Hideyuki
Enablers for Smart Cities

GIANNI Robert
Responsibility and Freedom (Responsible Research and Innovation Set – Volume 2)

GRUNWALD Armin
The Hermeneutic Side of Responsible Research and Innovation (Responsible Research and Innovation Set – Volume 5)

LAGRAÑA Fernando
E-mail and Behavioral Changes: Uses and Misuses of Electronic Communications

LENOIR Virgil Cristian
Ethical Efficiency: Responsibility and Contingency (Responsible Research and Innovation Set – Volume 1)

MAESSCHALCK Marc
Reflexive Governance for Research and Innovative Knowledge
(Responsible Research and Innovation Set – Volume 6)

PELLÉ Sophie, REBER Bernard
From Ethical Review to Responsible Research and Innovation
(Responsible Research and Innovation Set – Volume 3)

REBER Bernard
Precautionary Principle, Pluralism and Deliberation: Sciences and Ethics
(Responsible Research and Innovation Set – Volume 4)

VENTRE Daniel
Information Warfare – 2nd edition

Printed and bound by CPI Group (UK) Ltd, Croydon, CR0 4YY

15/09/2026

14962407-0002